Activity Based Costing
for Construction Companies

Activity Based Costing for Construction Companies

First Edition

Yong-Woo Kim
University of Washington,
Seattle, WA
USA

WILEY Blackwell

Registered Offices
John Wiley & Sons Ltd, The Atrium, Southern Gate, Chichester, West Sussex, PO19 8SQ, UK

Editorial Office
9600 Garsington Road, Oxford, OX4 2DQ, UK
For details of our global editorial offices, customer services, and more information about Wiley products visit us at www.wiley.com.

Wiley also publishes its books in a variety of electronic formats and by print-on-demand. Some content that appears in standard print versions of this book may not be available in other formats.

Limit of Liability/Disclaimer of Warranty

Library of Congress Cataloging-in-Publication data applied for

ISBN: 9781119194675

Cover image: Finbarr Carroll / EyeEm/Gettyimages
Cover design by Wiley

Set in 10.5/13.5pt Minion by SPi Global, Pondicherry, India
Printed and bound in Malaysia by Vivar Printing Sdn Bhd

10 9 8 7 6 5 4 3 2 1

Contents

Preface

It was the mid-1990s when I got my first project engineer job – my first job after I graduated from college. The first project was an industrial project (cement plant construction), where the general contractor I worked for experienced significant cost overruns, although all team members including labor crews worked hard. I learned that there was not much room for cost reduction since most work was subcontracted, with fixed price contracts. I did not learn how to effectively manage the activities of management staff and overhead costs, although I did learn how to effectively manage project direct costs, through the use of several tools such as the earned value method. Senior management at the project tried to reduce the number of management staff to reduce overhead costs, but it turned out that the lack of management staff caused more confusion and inefficiency on the site.

My experiences as a practitioner prompted me to study process improvement and management of overhead costs. As a graduate student of CAL, I was fortunate in two ways. First, I was very fortunate that I had chance to work with some of the great minds of Lean construction, a new construction management paradigm, which placed its focus on the production systems. Gaining knowledge and background in production systems helped me expand my horizons in studying overhead cost management. Second, I was fortunate that I had chance to study activity-based costing – a new cost management paradigm – in the manufacturing industry in the late 1990s, when activity-based costing had not yet been introduced in the construction industry.

However, the application of activity-based costing to the construction industry was not an easy job. Because of the difference between manufacturing and construction, namely the nature of the organizational structure and the production system, activity-based costing requires adaptation to suit the specifics of the construction industry. My understanding of the production system and organizational structure of the construction industry, and the principles of Lean construction, allowed me to apply activity-based costing to this new context.

My first ABC project was a commercial building project where ABC was applied to managing the general contractor's project overhead costs; this became my PhD dissertation in 2002. Since then, my ABC journey has continued. I expanded my ABC experience through consultation and research in different contexts, e.g., home office overhead costs, overhead costs at fabrication shops, etc.

I have noted the lack of reference materials on managing overhead costs in our industry. I felt the same way regarding construction education. It is true that the majority of cost management topics in the classroom in our industry still focus on managing direct costs such as earned value management and labor productivity. Although they are important, our industry and classroom need guidance on managing overhead costs. That is what motivated me to write this book.

This book has two main objectives: 1) to outline activity-based costing to be applied to different construction industry settings; and 2) to provide an implementation roadmap. I wanted to show the logic and simplicity behind ABC as well as the benefits of ABC. I hope that by reading the book, you will be able to create your own ideas about how to manage overhead costs in your organization.

I did not intend to write a complete treatment of an accounting method with which your accounting system is to be replaced. I did not write this book from an accountant's perspective. Instead, I wrote this book from the perspective of an operational manager who is concerned about their production system.

Once people recognize increased overhead costs, some tend to reduce them by reducing workforce or replacing existing staff with a less experienced workforce. It can be dangerous to reduce the number of management staff or to replace existing management staff with less experienced (i.e., less expensive) staff without fully examining the relationship between overhead costs and the production system. In this regard, the task of managing overhead costs should involve people who do understand the production system. That is the reason why this book was written for operational management staff.

I provide several case studies, each of which will give you an idea of how ABC can be implemented in a different setting. In each case, I tried to set out what I learned from my experience so that you can avoid pitfalls.

Activity-based costing is a powerful tool for managing overhead costs. I believe that the knowledge and experience of activity-based costing which I gained through past consulting and research experiences are limited and need improvement. However, the compelling reason I wrote this book is that any advancement begins with a foundation, and somebody needs to pave the way for practitioners and researchers in the domain of managing overhead costs.

Yong-Woo Kim, Ph.D.
University of Washington, Seattle

1

Introduction

Activity Based Costing for Construction Companies, First Edition. Yong-Woo Kim.
© 2017 John Wiley & Sons Ltd. Published 2017 by John Wiley & Sons Ltd.

Every business wants to reduce its costs so as to maximize its profits. Since construction is a type of business, it cannot be denied that every construction contractor is eager to reduce their costs. Construction contractors should also be able to accurately price each of their products and services (i.e., their projects), because accurate estimation of projects leads to the success of projects. Prior to addressing effectively managed costs, we need to have consensus on what comprises costs in a construction company.

1.1 What comprises costs in a construction company?

We usually define costs as a resource consumed to achieve a specific objective (Horngren *et al.*, 1999; Raffish and Turney, 1991). Costs are usually measured as the monetary amount that must be paid to acquire resources, i.e., goods and services.

Let us investigate the cost structure of a construction company (i.e., a general contractor) to establish what comprises a construction company's costs. The cost structure of a construction company is the framework by which its home offices and each of its projects are budgeted and controlled. Figure 1.1 shows the typical cost structure of a construction contractor whose revenue is the sum of the revenue of all projects.

Home office overhead (General overhead)		Employees
		Facility and utility
		Others
Construction costs (Project costs)	Project overhead costs	Employees
		Facility and utility
		Others
	Project direct costs	Direct labor
		Direct material
		Subcontractor
		Equipment

Figure 1.1   Cost structure of a construction contractor.

As seen in Figure 1.1, a contractor's total costs consist of total construction costs and its general overhead costs. Total construction costs are the sum total of the construction costs of each project, which includes project direct costs and project overhead costs. The terms "overhead" or "overhead costs" are used to represent indirect costs in the rest of this book.

1.1.1   Construction costs (project costs)

Construction costs include both direct construction costs and the overhead (indirect) costs of each project. Direct project costs are the cost of materials, labor, and equipment, and subcontract costs. They are consumed and incorporated into the construction costs of a specific project. Project overhead costs include the consumption of resources used to support the activities of direct construction costs (e.g., field jobs), such as the salaries of project engineers.

All construction costs should be charged to a specific construction project. In addition, some of the home office resources used by a specific project are considered to be part of construction costs (i.e., project overhead costs). Suppose that 50% of an LEED (Leadership in Energy and Environmental Design) engineer's time in charge of green construction consulting at your home office is spent on three construction projects. Then, 50% of his or her salary needs to be allocated to these three projects according to the actual percentage of time spent on each project. In other words, 50% of the LEED engineer's salary is considered to be project overhead costs.

1.1.2   Overhead costs in a construction company

Although we used several terms in relation to cost structure (Figure 1.1), in general, costs can be grouped into direct costs and overhead costs. There are multiple definitions for direct and overhead costs in construction. One definition of direct costs is the costs expended in the realization of a physical sub-element of the project (Halpin, 1985). Although some practitioners use this definition based on the realization of a physical element on site, the definition is not widely accepted in the domain of cost accounting.

A generally accepted definition of direct costs uses the ability to track a cost to a cost object.[1] Direct costs of a cost object are related to a particular cost object and can be traced to it in an economically feasible (cost-effective) way (Horngren *et al.*, 1999). The term "direct costs," when applied to construction accounting means costs which can be specifically identified with a construction job or with a unit of production within a job (Coombs and Palmer, 1989). This definition is consistent with the general definition of direct costs.

Overhead costs of a cost object, on the other hand, are related to a particular cost object but cannot be traced to it in an economically feasible way. Figure 1.2 illustrates cost categorization according to cost assignments. The term "cost allocation" is used to describe the assignment of overhead costs to a particular cost object (Raffish and Turney, 1991). The other important term is "cost object." According to the definition of overhead costs, two criteria for discerning overhead costs are (1) cost object and (2) traceability.

The term "overhead costs" is still used in a vague manner in the construction industry, because our industry has more than one type of overhead cost. In other words, the same cost can be both an overhead cost and a direct cost, depending on the perspective of the observer.

For example, a superintendent's wage is an overhead cost to a specific project at the project level, but it is also a direct cost to the

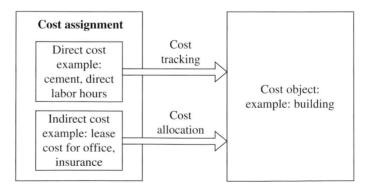

Figure 1.2   Cost assignment and classification.

---

[1] Cost object is defined as any product or service to which costs are assigned or tracked (Raffish and Turney, 1991).

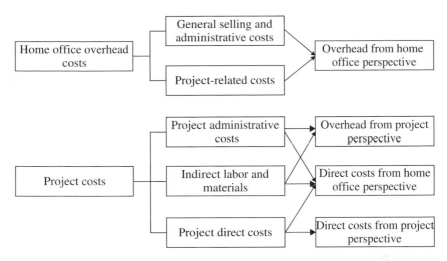

Figure 1.3   Cost classification of a construction contractor: duality of overhead costs.

specific project at the home office level. Figure 1.3 presents a general classification of construction costs showing the "duality" of overhead costs in the construction industry.

1.1.3   The cost classification in use and the duality of overhead costs

Despite definitions of cost classification, the construction industry does not seem to have a standard cost classification system of its own. Holland and Hobson's survey results (1999) suggested that cost categorization is not standardized in construction.

Table 1.1 shows the result of a survey of general contractors' cost categorization that was carried out by the authors' research team. In Table 1.1, project overhead costs (P) and project direct costs (D) constitute project construction costs, and overhead costs (O) refer to general overhead costs. Table 1.1 also confirms that the construction industry lacks a standard system for categorizing costs.

Why doesn't the industry have cost classification standards? The author's observations and interviews with industry professionals suggest that categorization is mainly driven by the terms of contracts. In other words, commercial interests bound by contracts lead to cost classification. Suppose that your contract says that the owner is to reimburse all of the project's direct costs.

Table 1.1   Cost classification examples in the construction industry (general contractor's perspective).

Cost Item No.	Item	Number of participants	P	D	O
1	Construction manager	15	60.0%	33.3%	6.7%
2	Superintendent	14	64.3%	28.6%	7.1%
3	Project engineer	15	60.0%	33.3%	6.7%
4	Material engineer	13	61.5%	30.8%	7.7%
5	Survey crew	12	66.7%	33.3%	0.0%
6	Quality staff	13	69.2%	23.1%	7.7%
7	Safety engineers	14	64.3%	28.6%	7.1%
8	Secretaries (field office)	13	69.2%	23.1%	7.7%
9	Automobiles	12	50.0%	33.3%	16.7%
10	Fencing and gates	11	81.8%	18.2%	0.0%
11	Temporary parking	13	61.5%	30.8%	7.7%
12	Project office	15	73.3%	20.0%	6.7%
13	Fabrication shop	9	33.3%	22.2%	44.5%
14	LEED engineer	8	37.5%	25.0%	37.5%
15	Rigging equipment	12	66.6%	16.7%	16.7%
16	Warehouse	14	50.0%	21.4%	28.6%

Note: P = project overhead costs, D = project direct costs, O = home office overhead costs.

In that case it is not surprising that you tend to include more cost items in the category of project direct costs. A lack of cost classification standards is one of the obstacles to the advancement of cost management practices.

1.2   Overhead costs in new business environments

Project overhead costs are increasingly important, as they have grown in recent years (Assaf *et al.*, 1999; Kim and Ballard, 2005). A number of driving forces have increased overhead costs (i.e., general overhead costs and project overhead costs) in recent years. The following four factors are identified as driving forces:

- Technical and managerial factor

  A fragmented approach and activity-centered management has caused ineffective project delivery (Ballard *et al.*, 2011). In response to such challenges, Lean construction and Building

Information Modeling (BIM) have been adopted widely in the construction industry. Technical and management innovations such as BIM and Lean construction increase a contractor's overhead costs, both in terms of project overhead costs and general overhead costs, although they contribute to reducing total construction costs (mostly direct costs).

The fundamental principle of Lean construction is a reliable work flow[2] through a predictable production plan (Ballard *et al.*, 2007; Ballard, 1999). Achieving a reliable work flow via Lean construction usually involves intensive collaborative production planning such as "pull" planning and weekly work planning (Ballard, 1994). Other Lean construction principles such as just-in-time (JIT) delivery also require intensive planning efforts and close collaboration among project stakeholders; however, allocating more resources to collaboration and planning has an impact on project overhead costs. In addition, a contractor needs to hold educational workshops or training sessions to educate employees in preparation for Lean implementation.

Building Information Modeling has been practiced in inter-organizational collaborations among architects, engineers, and construction contractors (Dossick and Neff, 2011). In addition to the hardware and software expenses that arise with this system, a construction company will need more staff to operate the system, as well as having to meet the costs of associated collaborative processes, such as coordination meetings (Eastman *et al.*, 2008).

In addition to using the Lean and BIM systems, the construction industry has also adopted information technology in many areas. Some examples of technology include ERP (enterprise resource planning), various material tracking systems such as bar codes and RFID (radio frequency identification), and automation. These investments in information technology inevitably increase overhead costs, despite increased management efficiency and labor productivity.

- Social factor

    The concept of green building or sustainable construction has emerged across the industry as society's awareness of

---

[2] The movement of information and materials through networks of interdependent specialists (Lean Construction Institute, 2016).

sustainability issues has grown. The construction industry, one of the nation's largest industries, is making an effort to reduce its environmental impact by adopting sustainable design and construction practices (Bae and Kim, 2007). The exemplary practice widely adopted in the industry is LEED (Leadership in Energy and Environmental Design).[3] When LEED is implemented in a project, it impacts not only direct costs (e.g., environmentally friendly materials) but also overhead costs. Examples of overhead costs incurred due to LEED may include professional fees and documentation costs (General Service Administration, 2004). Those soft costs (overhead costs) can be absorbed either into project overhead costs or into general overhead costs, depending on the attributes of each specific cost.

Additional overhead costs due to LEED certification will vary depending on the project type and context. Administration of the LEED certification process and documentation of LEED credits is an added cost directly associated with LEED certification. The USGBC estimates that these administration costs range from $20,000 to $60,000, depending upon project size, complexity, and project team experience (USGBC, 2009). Although various case studies on the cost of LEED exist, it is clear that society's awareness around green and sustainability issues increases overhead costs in our industry.

- Contractual factor

  The fragmented approach of procurement systems for construction projects, known as design-bid-build, has been a dominant delivery method. However, such a fragmented procurement approach has affected project effectiveness in that it does not encourage integration, collaboration, and communication among organizations participating on a project (Love *et al.*, 1998). In particular, decoupling of design and construction has exacerbated communication and collaboration problems in complex and uncertain projects (Ballard *et al.*, 2011).

  In response to the challenges caused by fragmented procurement, alternative delivery systems have been actively

---

[3] LEED is a certification program supporting sustainable design and construction practices developed and endorsed by the non-profit U.S. Green Building Council (USGBC) in the building construction industry. (USGBC, 2009).

adopted both in the private market and in the public market. Alternative delivery systems include (1) CM/GC[4] or CM-at-Risk,[5] (2) Design-Build, and (3) the Integrated Project Delivery System.

The Design-Build Institute of America (DBIA) published a report on the use of the design-build delivery method in the United States in 2013 (Design-Build Institute of America, 2013). The study reports that in 2010, 41% of non-residential building projects in the United States were delivered using design-build, an increase from 30% in 2005. The same report shows that the use of design-bid-build declined from 67% in 2005 to 53% in 2010. The number of projects built with the CM-at-Risk system grew from 3% in 2005 to 6% in 2010 (Design-Build Institute of America, 2013).

The use of alternative delivery systems inevitably increases overhead costs, for the following reasons:

1) The involvement of more contractors in the design phase (or preconstruction services) increases the need for resources in preconstruction services, such as design professionals or preconstruction managers.
2) The negotiated contract, as an alternative to low-bid contractor selection, requires marketing resources to build relationships with potential customers.

In addition to the various factors which increase a contractor's overhead costs, we need to pay attention to the construction environment, where the use of specialty contractors has increased (Lock, 2000). As part of this trend, general contractors have lost control of management of project direct costs on projects where all the work divisions are subcontracted. As compared to projects where general contractors use directly hired work in most work divisions, more coordination is required to manage various specialty contractors, each with its own commercial interests. As a result, learning to effectively manage overhead costs (both general overhead costs and project overhead costs) is key to making a general contractor competitive in the market.

---

[4] Construction Management/General Contrator Delivery Method.
[5] Construction Management-at-Risk.

1.3   Role of overhead cost management

Most guidelines for managing overhead costs are for tax, financial reporting, and claim purposes (Coombs and Palmer, 1989). How overhead costs are allocated affects financial reporting and tax accounting (Coombs and Palmer, 1989). In fact, allocating overhead costs to different sectors or projects is a way of maneuvering a company's taxable profits. Controlling overhead costs from those perspectives is important, but overhead cost control from a managerial perspective has rarely been studied.

In the previous section, we discussed the importance of managing overhead costs in current construction environments. In addition to those issues, it must be asked: what is the purpose of managing overhead costs in a construction firm?

1.3.1   Overhead costing system should provide accurate costing on cost objects

One of the fundamental functions in managing overhead costs is to accurately allocate them to the proper cost objects. Cost objects can vary depending on the purposes of the costing system. As a contractor, your cost object can be each project when allocating your general overhead costs. Each work division on a specific project can be a cost object if you want to investigate the efficiency of management activities at the project level.

Accurate allocation of general overhead costs to projects is important for two reasons. First, accurate allocation of general overhead costs provides a general contractor with accurate information on profits for each project. Accurate profit information on cost objects allows contractors to identify where they lose or make money through accurate allocation of overhead costs. In addition, accurate profit information gives a general contractor insight into marketing strategies that focus on profitable project sectors. In negotiated contracts, such information becomes important in building and nurturing relationships with potential customers, as compared to low-bid contracts where such information is not as important.

Second, accurate cost information on each project enables a general contractor to estimate costs on future projects more accurately, putting him or her in a superior position in bidding for a job. The current practice of applying a predetermined percentage to all types of projects may lead to inaccurate cost estimation, which can make the contractor less competitive in the market. Although some contractors use different ratios for general overhead costs on different types of projects when bidding, inaccurate allocation of overhead costs makes bidding numbers inaccurate, thereby making them less competitive in the market.

1.3.2   Overhead costing system should contribute to reducing total costs without sacrificing value

As discussed, assuming that all work divisions are subcontracted using a fixed-cost contract, there is not much room to reduce construction direct costs. However, reducing overhead costs is not always a great way to maximize profits. The fundamental role of overhead costs (i.e., the consumption of overhead resources) is to effectively support fieldwork on sites. Overhead costs include scheduling, estimating, inspection, supervising fieldworks, safety training, etc.

These are support activities that enable fieldwork, although they are not directly value-adding. For example, production planning or weekly scheduling requires multiple activities, each consuming overhead resources. The value delivered by this consumption of overhead resources is to give field crews job assignments. It is necessary to understand activities, as well as the values associated with the consumption of overhead resources, before tackling the challenge of reducing some of your overhead costs.

The key is to understand activities which consume overhead resources and the value of the activities performed. You should be careful that the value of activities associated with overhead costs does not get lost as you try to reduce your overhead costs. Your overhead costing system needs to help you to achieve a reduction in overhead costs without sacrificing the value that those costs generate.

1.4   Structure of this book

This book is organized in six chapters, including this one. Chapter 2 describes the problems of current cost accounting methods for managing overhead costs and the methodology of activity-based costing (ABC). Chapters 3, 4, and 5 deal with various cases to which ABC could be applied. Chapter 6 provides guideline for your ABC implementation.

References

Assaf, S., Bubshait, A., Atiyah, S., and Al-Shehri, M. (1999). "Project overhead costs in Saudi Arabia," *Cost Engineering*, 41(4), 33–38.

Bae, J. and Kim, Y. (2007). "Sustainable value on construction projects and lean construction," *Journal of Green Building*, 3(1), 156–167.

Ballard, G. (1994). "The last planner," *Northern California Construction Institute Spring Conference*, Monterey, California.

Ballard, G. (1999). "Improving work flow reliability," in *Proceedings of the 7th Annual Conference of the International Group for Lean Construction*, Tommelein, I. D. (editor), 275–286.

Ballard, G., Kim, Y., Azari, R., and Cho, S. (2011). *Starting from Scratch: A New Project Delivery Paradigm*, Construction Industry Institute, Research Report 271, Austin, TX.

Ballard, G., Kim, Y., Min, L., and Jang, J. (2007). *Lean Implementation at a Project Level*, Construction Industry Institute, Research Report 234, Austin, TX.

Coombs, W. and Palmer, W. (1989). *Construction Accounting and Financial Management*, 4th edn, McGraw-Hill, Inc., New York, NY.

Design-Build Institute of America (2013). *Design-Build Project Delivery Market Share and Market Size Report*, Washington, DC.

Dossick, C. and Neff, G. (2011). "Messy talk and clean technology: communication, problem-solving and collaboration using Building Information Modeling," *The Engineering Project Organization Journal*, 1(2), 83–93.

Eastman, C., Teicholz, P., Sacks, R., and Liston, K. (2008). *BIM Handbook: A Guide to Building Information Modelling for Owners, Managers, Designers, Engineers, and Contractors*, John Wiley and Sons Ltd, New Jersey.

General Service Administration (2004). *LEED Cost Study Final Report*, Contract No. GS–11P–99–MAD–0565, Washington, DC.

Halpin, D. (1985). *Financial & Cost Concepts for Construction Management*, John Wiley and Sons Ltd, New York, NY.

Holland, N. and Hobson, D. (1999). "Indirect cost categorization and allocation by construction contractors," *Journal of Architectural Engineering*, ASCE, 5(2), 49–56.

Horngren, C., Foster, G., and Datar, S. (1999). *Cost Accounting*, 10th edn, Prentice Hall, Upper Saddle River, NJ.

Kim, Y. and Ballard, G. (2005). "Profit-point analysis: A tool for general contractors to measure and compare costs of management time expended on different subcontractors," *Canadian Journal of Civil Engineering*, 32(4), 712–718.

Lock, D. (2000). *Project Management*, 7th edn, Gower, Vermont, VM.

Love, P., Gunasekaran, A., and Li, H. (1998). "Concurrent engineering: a strategy for procuring construction projects," *International Journal of Project Management*, 16(6), 375–383.

Raffish, N. and Turney, P. (1991). "Glossary of activity-based management," *Cost Management Journal*, Fall, 53–63.

U.S. Green Building Council (2009). *Green Building Design and Construction*, U.S. Green Building Council, Washington, DC, 645 pp.

2 What Is Activity-Based Costing?

Activity Based Costing for Construction Companies, First Edition. Yong-Woo Kim.
© 2017 John Wiley & Sons Ltd. Published 2017 by John Wiley & Sons Ltd.

Every construction company is eager to manage its overhead costs. What then does "managing overhead costs" mean? At the very least, it requires a company to track and compile its overhead costs. Tracking and compiling overhead costs is required not only for managing overhead costs, but also for calculating costs for its products or services. In project-based industries such as the construction industry and the ship-building industry, a job costing method is commonly used. At the heart of construction project accounting is the job costing system. In the job costing system, a cost object is an individual unit, batch, or lot of a distinct product or service called a job (Horngren *et al.* 1999).

It makes sense that resources are to be assigned to cost objects. The issue is *how* resources are assigned to cost objects – a process which defines an accounting method. This chapter first reviews the current method prevailing in the industry. It then addresses the problems of the current method and follows with a methodology of activity-based costing.

2.1   Traditional accounting method: resource-based costing with volume-based allocation

2.1.1   *Resource-based costing*

The premise of traditional cost accounting is that products or services consume resources directly; this is called "one-stage costing," as shown in Figure 2.1 (Miller, 1996). One-stage costing allocates the overhead resources to products or services.

The essential attribute of traditional overhead costing based on the above assumption is the use of a single cost pool and a single overhead cost rate to allocate overhead costs to cost objects (Cokins, 1996; Miller, 1992; Horngren *et al.*, 1999).

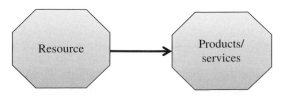

Figure 2.1   One-stage costing.

(a)

Job	Description	Costs
	direct cost	
10	Form, foundation building 01	$ 11,000.00
20	Form, foundation building 02	$ 6,000.00
30	Form, foundation building 03	$ 3,800.00
40	Rebar, foundation building 01	$ 10,400.00
50	Rebar, foundation building 02	$ 5,200.00
60	Rebar, foundation building 03	$ 3,800.00
	Subtotal	$ 40,200.00
160	Supervisor (1)	$ 5,500.00
170	Project engineer (2)	$ 9,000.00
180	Project manager (1)	$ 7,500.00
190	Warehouse guard (1)	$ 3,500.00
200	Helper (2)	$ 4,000.00
	Subtotal	$ 29,500.00
	Total	$ 69,700.00

(b)

Job	Description	Costs
10	Form, foundation building 01 (Material)	$ 3,000.00
20	Form, foundation building 02 (Material)	$ 2,000.00
30	Form, foundation building 03 (Material)	$ 1,500.00
40	Rebar, foundation building 01 (Material)	$ 8,000.00
50	Rebar, foundation building 02 (Material)	$ 4,000.00
60	Rebar, foundation building 03 (Material)	$ 3,000.00
	Subtotal	$ 21,500.00
100	Form, foundation building 01 (Labor)	$ 8,000.00
110	Form, foundation building 02 (Labor)	$ 4,000.00
120	Form, foundation building 03 (Labor)	$ 2,300.00
130	Rebar, foundation building 01 (Labor)	$ 2,400.00
140	Rebar, foundation building 02 (Labor)	$ 1,200.00
150	Rebar, foundation building 03 (Labor)	$ 800.00
	Subtotal	$ 18,700.00
160	Supervisor (1)	$ 5,500.00
170	Project engineer (2)	$ 9,000.00
180	Project manager (1)	$ 7,500.00
190	Warehouse guard (1)	$ 3,500.00
200	Helper (2)	$ 4,000.00
	Subtotal	$ 29,500.00
	Total	$ 69,700.00

Figure 2.2   Example of cost reports (Kim and Ballard, 2001).

Consider a simple case where a building project includes construction of three different buildings. Figure 2.2 shows the information that the current method can provide. Costs are categorized into three types of resources: labor, material, and management. Figure 2.2(a) shows the information when direct material costs and direct labor costs are integrated into a category of direct cost, whereas Figure 2.2(b) shows the information when they are presented separately.

Accounting information on overhead costs is shown in accounts 160 through 200. As seen in Figure 2.2, this accounting information (accounts 160 through 200) shows how much of each resource is consumed, but it does not present any information on activities or processes. In other words, the current method doses not provide a process view because costs are categorized in terms of resources rather than activities or processes.

2.1.2   Overhead costs allocation

Suppose that three buildings (building #1, #2, and #3) are cost objects. Does a contractor need to allocate its project overhead costs to three cost objects? The answer should be "it depends on the needs." In my observation, overhead costs are not usually assigned to each building. However, if a contractor wants accurate cost information that can be used as a reference in future bidding, it is a different story. If it is necessary to allocate overhead costs to

Table 2.1   Overhead assignment example.

	Building #1	Building #2	Building #3	Total
Direct material	$11,000.00	$6,000.00	$4,500.00	$21,500.00
Direct labor	$10,400.00	$5,200.00	$3,100.00	$18,700.00
Total direct costs (3) = (1) + (2)	$21,400.00	$11,200.00	$7,600.00	$40,200.00
Total overhead				$29,500.00
Assignment % (labor costs)	55.61%	27.81%	16.58%	
Overhead (4)	$16,406.42	$8,203.21	$4,890.37	
Total (3) + (4)	$37,806.42	$19,403.21	$12,490.37	

Notes:
- Three buildings (buildings #1, #2, and #3) are cost objects.
- Tracking direct costs (i.e., direct material and direct labor) to each building is not difficult. The problem is how to allocate overhead costs to each cost object.
- Direct costs shown (rows 1 and 2) allocated to each building are arbitrary, not derived by any calculation.

different buildings, then the question should be how overhead costs are to be allocated fairly to different cost objects (i.e., three different buildings in this case). In general, construction contractors use a kind of single allocation base in allocating their overhead costs. The allocation bases that are commonly used in the industry include:

- direct labor costs
- direct labor hours
- revenue (or contract amount).

Suppose that a contractor uses its direct labor costs as an allocation base. The cost allocation result using direct labor costs as an allocation base is to be found in Table 2.1. As shown in Table 2.1, total overhead costs of $29,500 are assigned in the following percentages: 55.61% ($10,400/$18,700), 27.81% ($5,200/$18,700), and 16.58% ($3,100/$18,700).

2.2   What are the problems with the current method?

You are probably already used to the current method (i.e., resource-based costing with volume-based allocation), and perhaps you do not see any critical problems with it. Let us revisit the previous chapter where two important roles of the overhead cost management system are addressed.

Role of overhead cost management

1) The overhead costing system should provide accurate pricing/costing so that users can identify where they lose or make money by reasonable allocation to cost objects.
2) The overhead costing system should contribute to reducing or minimizing total costs.

2.2.1 Is the current method contributing to reducing total costs?

Does the current method provide information to help users reduce total costs? Because cost per resource is the only type of information available, users may need to reduce resources with high overhead costs. However, users may not want to reduce such resources simply because they cost more than others. Figure 2.2, for example, indicates that project engineers cost more than other resources, perhaps leading to a desire to reduce their use. It may not be desirable, however, to reduce the cost of project engineers, although their cost ratio (project engineer costs to total overhead costs) may be higher for a particular project than for other projects. Usually, such decisions about where to cut costs require cost information on activities consuming the resources.

Because the traditional costing method assumes that resources are directly consumed by products or services, it lacks the ability to provide proper process visibility. Invisibility as to where the costs accumulate conceals process waste (Cokins, 1996; Miller, 1996).

2.2.2 Does the current method provide accurate pricing?

In the examples above (Figure 2.2 and Table 2.1), we can easily observe that the method of resource-based costing is valid only if the value of direct labor costs is the critical factor in determining the quantity of resources consumed by overhead activities. If building #3 consumes more overhead resources because the building requires more design changes or administration than other buildings, building #3 will be under-costed within the volume-based allocation method. The example shows that the overhead cost allocation using direct labor costs is invalid.

Managerial accountants have become aware of the declining relevance of the numbers they produce for end users and decision makers (Johnson and Kaplan, 1987; Cokins, 1996). Since the late 1980s, many professionals and researchers have criticized traditional cost accounting. Johnson and Kaplan (1987) criticized the traditional managerial cost information as too distorted to be useful for management planning and control decisions. In general, the traditional managerial accounting system fails to provide accurate product costs (Johnson and Kaplan, 1987; Horngren *et al.*, 1999).

2.3   What is activity-based costing?

2.3.1   *Definition*

A new accounting method, activity-based costing (ABC), has been developed as a means of overcoming the problems of traditional cost accounting, mainly (1) lack of contribution to reduction of total costs and (2) cost distortion due to misallocation of overhead costs. Activity-based costing is an accounting methodology that identifies activities in a production system or an organization (Johnson, 1992). ABC assigns the costs of each activity to cost objects in proportion to the actual consumption of activities by each cost object. ABC is not a replacement for general ledger accounting but a translator inserted to extract general ledger and other data (Cokins, 1996). As a metaphor for ABC, Cokins (1996) suggested a model of a corrective optical lens that brings clarity by reassigning costs.

The following two propositions are central to ABC (Miller, 96):

1)   Activities consume resources. (Resources are assigned to activities.)
2)   Products consume activities. (Activities are assigned to cost objects based on their usage.)

Figure 2.3 shows the information flow of cost assignment. While the traditional costing method only addresses the costs of resources and assigns resource costs to cost objects directly, ABC

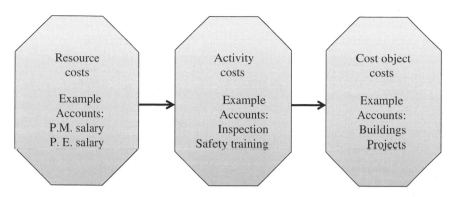

Figure 2.3   Information flow in activity-based costing.

goes one step further, estimating costs for activities. The process of assigning resource costs to activities and activity costs to cost objects requires a method of cost tracking or reasonable allocation. Usually, assigning resource costs to activities is a process of directly matching the consumption of resources with each activity (Kaplan and Cooper, 1997). On the other hand, the process of assigning activity costs to cost objects requires cost allocation, which is a process of applying activity costs to cost objects using an activity cost driver. Cost tracking and cost allocation are addressed in the following chapters.

2.3.2   *Characteristics of ABC*

An ABC system is different from traditional costing systems in many ways. The following major characteristics of ABC differentiate it from the traditional cost accounting system.

2.3.2.1   ABC uses two-stage costing

As mentioned in the previous sections, traditional cost accounting uses "one-stage costing" whereby resources are directly assigned to cost objects such as projects or buildings using volume-based allocation (e.g., contract amounts or direct labor costs). The ABC system provides an expanded set of potential cost drivers. While traditional costing systems use volume or volume-based allocation as a cost driver, ABC uses two different types of drivers: resource (cost) drivers and activity cost drivers. In addition, there are three types of activity cost drivers (Cooper and Kaplan, 1992).

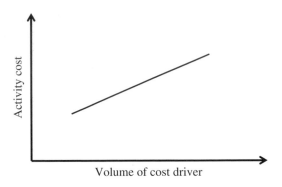

Figure 2.4 Activity cost driver vs. activity costs.

General ledger accounts (i.e., resource costing) are tracked to various activities in an appropriate proportion using resource cost drivers (e.g., time or area). Activity costs are then distributed to final cost objects using activity cost drivers. An activity cost driver can be defined as any factor that causes a change in the consumption of an activity by other products, suppliers, or customers (Cokins, 1996; Raffish and Turney, 1991). In effect, an activity cost driver refers to any factor that shows a linear relationship with an activity cost. An activity cost driver is also a factor the volume of which increases as an activity cost increases, as shown in Figure 2.4.

2.3.2.2 The ABC system recognizes different types of activities (hierarchy)

ABC recognizes four different levels of activities, a hierarchy that maintains that costs are driven by, and are variable respective to, activities that occur at different levels. Because construction activities occur in a different context than manufacturing activities, activities in construction have a different hierarchy. In this section, we will first review a hierarchy of manufacturing activities, then we will discuss this in the construction context.

In manufacturing, the ABC system usually has four different levels of activities: output unit level, batch level, product-sustaining level, and facility-sustaining level, as shown in Figure 2.5 (Cooper, 1989; Cooper and Kaplan, 1992; Horngren *et al.*, 1999). Output unit-level costs are defined as resources consumed on activities performed on each individual unit of a product or service. Manufacturing operating costs such as energy and repair that are

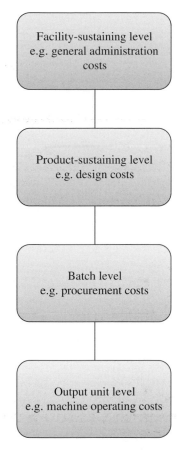

Figure 2.5   Activity hierarchy in manufacturing (Horngren, 2000).

related to the activity of running machines are output unit-level costs. Batch-level costs are defined as resources consumed on activities that are related to a group of product units or services rather than to an individual unit of product or service. Setup and procurement costs are examples of batch-level costs in manufacturing. Product-sustaining costs are defined as resources consumed on activities undertaken to support individual products or services. Design costs and engineering costs are examples of product-sustaining costs in manufacturing. Facility-sustaining costs are defined as resources consumed on activities that cannot be tracked to individual products or services, while supporting the organization as a whole.

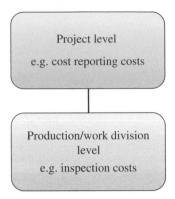

Figure 2.6   Activity hierarchy in construction project overhead costs.

In construction, ABC systems commonly take advantage of a cost hierarchy with different levels of activities: production system (or work division) level, project level, and organization level. As discussed in Chapter 1, construction companies usually have two types of overhead costs: project overhead costs and general overhead costs. Each type of overhead costs has a different hierarchical structure of activities.

Project overhead costs have two levels of activities as shown in Figure 2.6: production system (or work division) level and project level. Production system-level (or work division-level) costs are defined as project overhead resources consumed by activities performed on each task or work division (e.g., concrete, mason, or mechanical work). The costs of inspection activities in relation to specific tasks or work divisions are an example of costs at the production system level. Project-level costs are defined as project overhead resources consumed by activities undertaken to support a project. Costs of "project cost reporting" or "training activities" are examples of costs at the project level.

General overhead costs have two levels of activities as shown in Figure 2.7: project level and organization level. Project-level costs are defined as home office resources used for activities performed on a specific project. Costs of engineering activities or of sustainability-related activities to support a project's green building accreditation are examples of costs at the project level. Organization-level costs are defined as home office resources used for activities for sustaining a company. Costs of recruiting or costs of customer relationship development activities are examples of project-level costs.

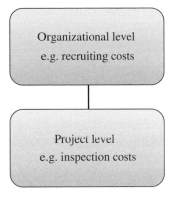

Figure 2.7   Activity hierarchy in construction general overhead costs.

2.3.3   Objectives of ABC system

An ABC system can be implemented with varied objectives depending on how the system is constructed. However, there are two generally accepted goals in most ABC systems.

2.3.3.1   To remove cost distortions

The traditional costing method leads to cost distortion with a one-stage costing (i.e., resources are directly consumed by cost objects). ABC can prevent cost distortion by using appropriate activity cost drivers rather than volume-based cost drivers (e.g., direct labor costs, contract amount) (Horngren *et al.*, 1999; Cokins, 1996; Miller, 1996). Through accurate allocation of overhead costs, the process of estimating costs of cost objects becomes transparent and accurate. Managers can use accurate cost information on cost objects for the purpose of strategic planning or operational management. Ways of using ABC cost information will be discussed in the following chapters.

2.3.3.2   To help eliminate or minimize low-value-adding activities

The second objective is to identify low-value-adding activities. ABC system development usually requires activity analysis through which activities in each department are identified, and resource consumption on each activity is investigated (Brimson, 1998). From activity analysis results, managers can identify activities that do not add value or consume resources to a great extent. In some cases, each activity can be categorized into different levels such as value-adding or non-value-adding for activity management purposes. For this reason, the ABC system is said to contribute to minimizing or reducing low-value-adding or non-value-adding activities. In this regard, ABC has similarities to Lean concepts (see 'Extension' below) (Cokins, 1996; Kim, 2002; Miller, 1992; Miller, 1996).

The main general objectives of abc systems

1) To provide accurate costing by removing cost distortions.
2) To help identify low-value-adding activities.

Extension. Lean construction

Lean construction comes from recognizing the limitations of current project management and applying new production management or "lean production" to the construction industry. Lean construction is defined as a production management-based approach to project delivery (Ballard *et al.*, 2007). Koskela (1992) criticized construction project management for modeling construction as a series of conversion activities. He argued that construction should adopt the new production philosophy, which improves competitiveness by identifying and eliminating waste (non-value-adding activities). His proposal that the construction industry should adopt the new production philosophy gave birth to "Lean construction."

2.4   Implementing activity-based costing

As the objectives of ABC system can vary, so too can the ways in which ABC is implemented. In this section, the general process of ABC implementation is described, while detailed methods for specific cases will be discussed in later chapters.

2.4.1   *Develop an activity-based costing charter*

I frequently mention in ABC seminars that the success of implementing activity-based costing is highly dependent on the amount of up-front planning and the degree of consensus from system development participants. One piece of rudimentary advice is to generate a team charter as soon as a task force is formed. A team charter serves many purposes: planning, communication, and building consensus. It is imperative to get consensus from key organization members because data collection is heavily dependent on surveys or interviews in most cases. In your charter, you can include whatever elements you want. However, the following are basic elements you should include in your team charter.

2.4.1.1   Objectives

This section defines what specific purpose(s) you want to achieve through your ABC system. A clearly defined set of objectives helps you design your ABC system, and includes the defined scope of the ABC system and the methods of data collection and analysis.

2.4.1.2   Scope

This section defines the types of costs in the cost structure to be investigated. The scope of your data collection should be clearly defined; a clearly defined scope also helps you to ensure that the right people are included in the team. Departments to be investigated would be determined according to the defined scope. A well-defined scope eventually reduces the risk that the team will spend valuable time on unnecessary data.

2.4.2   Define cost objects

A cost object is a product or service for which costs are measured, tracked, or allocated. Cost objects are also the output of an ABC system, being entities in respect of which activities are performed. Typical cost objects in ABC systems include products, services, projects, or customers.

Cost objects can vary depending on two major factors.

First, cost objects can vary depending on the *scope* of the ABC system. For example, in a construction project, the cost objects can be the individual buildings, work divisions, or subcontractors. On the other hand, the cost objects could be individual projects, services, or customers on a home office level.

Second, cost objects can vary depending on the *purpose* of the ABC system. On an organizational level (home office overhead), for example, cost objects can be individual customers if the company wants to identify profitable customers that they want to focus their marketing efforts on. Types of projects (i.e., residential, commercial, and heavy civil projects) can be cost objects if the company wants to identify profitable and unprofitable types of projects.

Typical cost objects

Organization level (home office overhead)
- Individual projects
- Customers

Project level (project overhead)
- Individual facilities
- Work divisions

2.4.3 *Identify activities*

Once you identify cost objects, you need to identify activities. An activity can be defined as a job consuming resources to achieve a specific goal. In practice, you can identify a large number of activities performed to produce outputs. Although it is hard to define the size of an activity quantitatively, an activity is a concept larger than a daily task, but smaller than a process or function.

Daily task < Activity < Function

For example, consider the process of budget approval in a company's home office. This process can be deconstructed into numerous sub-activities: checking quantity-takeoff, checking unit price, checking a project schedule, etc. For such sub-activities representing a detailed process description, it is hardly useful to track and calculate resource consumption and allocate their costs to cost objects, albeit that decomposing activities into detailed processes is useful in process improvement activities such as value stream mapping. If you define activities at a detailed process level, your ABC system grows too complex to maintain and update.

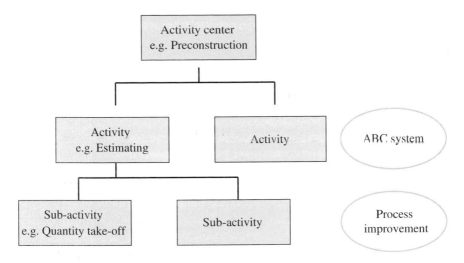

Figure 2.8   Level of detail in activity definition.

Figure 2.8 shows the level of activity detail. In practice, many organizations tend to identify their departments or functional units (e.g., green building unit) as activity centers. Each activity center has several activities, each of which may have sub-activities.

Activities are usually identified through interviews with managers such as department heads and supervisors. Interviewees are asked to identify major activities. The interview process is most effective if the interviewees are informed of the structure of activities, such as shown in Figure 2.8, prior to the interview. Interviews can last between 30 minutes and two hours.

2.4.4   *Assign resource costs to activities*

After activities are identified, resource costs need to be assigned to activities and activity centers. Resources include indirect labor (human resources), material and supplies, facilities (building space), equipment, and utilities. In a simple system, the scope of resources can be limited to indirect labor.

The cost of resources can be assigned to activity centers and activities in two ways: direct tracking and reasonable allocation. In a direct tracking method, the actual usage of resources by activities must be measured. Indirect labor or overhead workforce used to implement an activity can be directly measured in hours and charged to the activity.

$$C_i = \sum_{j=1}^{N} (UR_j \times Q_{ij})$$ [Equation 2.1]

Where, i = Activity number
    j = Resource number
    C = Cost
    UR_j = Unit rate of Resource $_j$
    Q_{ij} = Quantity of Resource j consumed to implement Activity $_i$

For example, persons performing activities can record time spent on each activity (i.e., time-tracking). Consider the case where an activity of budget approval needs the following resources per month on average:

According to Equation 2.1, the cost of budget approval can be calculated as follows:

C (budget approval) = $65 × 35 + $85 × 25 + $100 × 10 = $5,400

Although the direct tracking method provides accurate costs, it is expensive because it requires measurement of the actual usage. In addition, measurement of resource consumption other than human resources is hard to capture. When providing consulting services, I usually do not recommend using the direct tracking method unless it is needed for special purposes (e.g., claims against clients).

The method of *reasonable* resource allocation is preferred by most organizations in calculating the cost of activities. Surveys are commonly used for this purpose. Table 2.3 shows the time–effort percentage method in which the unit price of each resource (Table 2.2) and percentage of each employee's time–effort spent on each activity are used to calculate the cost of each activity. For example, the cost of Activity #1 can be calculated as

Cost of Activity #1 = 20% × $60/hour × 40 hours/week × 4 weeks (employee #1) + 30% × $80/hour × 40 hours/week × 4 weeks (employee #3) + ... = $17,760

Table 2.2   Example of time data on budget approval.

	Time consumed (hours)	Unit rate (hourly)
Engineer #1	35	$65
Engineer #2	25	$85
Engineer #3	10	$100

Table 2.3   Example of time–effort percentage method.

	Unit rate (per hour)	Activity #1	Activity #2	Activity #3	Activity #4	Total
Employee #1	$60	20%		30%	50%	100%
Employee #2	$75		40%		60%	100%
Employee #3	$80	30%	40%	30%		100%
...	...	...	...	...	...	...
		$17,760	$18,784	$26,720	$23,296	$86,560

Table 2.4   Example of facility costs allocated to departments.

Facility Costs = $200,000			
	Areas (SF)	Area ratio	Facility costs allocated ($)
Department #1	2,200	0.24	48,888.89
Department #2	1,500	0.17	33,333.33
Department #3	2,300	0.26	51,111.11
Department #4	1,200	0.13	26,666.67
Department #5	1,800	0.20	40,000.00
Total	9,000	1.00	200,000.00

If overhead costs include resources other than human resources, you may need a different method. First of all, you need to choose whether other overhead costs such as facilities (e.g., building ownership or rents) or utilities are included in your ABC system. If you want to include them in your ABC system, they can be allocated to activities as follows.

First, facilities and utilities costs can be allocated to different departments based on the area that each department occupies. Second, departmental costs (or functional costs) can be allocated to activities based on indirect labor costs ratio.

Below is an example of allocation of facilities costs ($200,000). A total facility cost of $200,000 is allocated to five different departments based on the area occupied by each department, as shown in Table 2.4. For example, Department #5 occupies 1,800 square feet, 0.2 ratio (1,800/9,000); then, $40,000 facility cost is allocated to Department #5 (200,000 x 0.2 = 40,000).

Suppose that Department #5 has four activities, as shown in Figure 2.8. The facilities cost of $40,000 allocated to Department #5 is allocated to four activities based on the ratio of activity costs as shown in Table 2.5.

Table 2.5   Example of facility costs (Department #5) allocated to activities.

Facility Costs (Department #5) = $40,000			
	Activity costs	Activity cost ratio	Facility costs allocated ($)
Activity #1	17,760	0.20	8,207.02
Activity #2	18,784	0.22	8,680.22
Activity #3	26,720	0.31	12,347.50
Activity #4	23,296	0.27	10,765.25
Total	86,560	1	40,000.00

Methods of allocating resource costs to activities
1) Direct tracking 2) Reasonable allocation   Examples: area ratios, time–effort percentage method

2.4.5   Assign activity costs to cost objects

After the costs of activities are calculated, the next step is to assign activity costs to cost objects. Direct tracking of every activity to cost objects is not economically feasible. Instead, companies need to link activity costs and cost objects. The link between activity costs and cost objects is called a "cost driver." A cost driver refers to a factor that is directly proportional to the costs of an activity. For example, the number of purchase orders can be a cost driver for an activity of "purchasing." The cost of cost objects should be the product of a unit activity cost and the volume of a cost driver (Equation 2.2).

$$C_k = \sum_{i=1}^{N}(UR_i \times Q_i) \qquad \text{[Equation 2.2]}$$

Where, k = Cost object number
i = Activity number
C = Cost
UR_i = Unit Rate of Activity $_i$
Q_i = Quantity (Volume) of Cost Driver for Activity $_i$

There are three types of cost drivers: transactional drivers, duration drivers, and budget drivers. Transactional drivers are the most common in ABC systems. Transactional drivers capture the number of times an activity is performed. They assume that the quantity of resources consumed to perform an activity each time remains constant. For example, the activities of subcontract revision review and its approval may use the number of revisions as a cost driver. If your company had 25 subcontract revisions and unit price per revision is assumed to be $540, then the cost of subcontract revision approval can be calculated:

$$\$540 \times 25 = \$13,500$$

Duration drivers measure the length of time over which an activity is performed. It is assumed that the resources are consumed with significant variations depending on when the activities are performed. For example, we cannot rely on the number of weekly reports as a cost driver to calculate the costs of the activity of "weekly reporting" if the amount of time that is required to finish the same activity varies widely from time to time (the same type of activity can take anywhere from 30 minutes to four hours). Since there is typically high variance in the amount of time expended on construction overhead activities, it is generally not appropriate to use counts of activity instances as the driver.

Table 2.6 summarizes general recommendations on how to choose the type of cost drivers to be used based on the consistency of resource consumption by an activity. If an activity consumes resources with little variance, which is expressed as "consistent resource consumption" in Table 2.6, then the costs of cost objects can be measured using a transactional cost driver. In other words, the costs of cost objects can be estimated simply by counting how many times an activity is performed on each cost object. For example, for an activity of developing a weekly report consuming 2 hr ~ 2.2 hrs, the number of weekly reports can be used as a cost driver (transactional driver). Note that there are no quantitative criteria to distinguish the types of cost drivers.

Table 2.6   Cost driver selection criteria.

Consistency of resource consumption by an activity	
High	Low
Transactional cost driver	Duration cost driver or Budget cost driver

Table 2.7   Example of budget ratio.

	Concrete	Steel	Interior
Budget, material	$200,000	$300,000	$150,000
Budget ratio (%)	30.77%	46.15%	23.08%

Therefore, the choice of cost driver is usually highly dependent on the experience of system designers.

In addition to transactional and duration drivers, ABC can use budget drivers when allocating activity costs to cost objects. In using a budget driver, activity costs are allocated to cost objects based on the budget ratio of the cost objects. Companies use budget drivers or duration drivers when the resource usage of an activity is not consistent. However, budget drivers are preferred when tracking the time spent on the activity is not economically feasible and transactional cost drivers are not suitable. For example, an activity of material management reporting may use the ratio to total material budget for each work division, if a work division is the cost object. Suppose that the project consists of three work divisions: concrete, steel, and interior work. Information on budget for material is shown in Table 2.7.

Suppose that you want to use a budget cost driver in assigning the cost of an activity of material management reporting to three cost objects. Provided that the monthly cost of material management reporting is $25,000, the cost of material management reporting activity can be distributed to three work divisions as follows:

Activity costs allocated to concrete = $25,000 × 30.77% = $ 7,692.50
Activity costs allocated to steel = $25,000 × 46.15% = $ 11,537.50
Activity costs allocated to interior = $25,000 × 23.08% = $ 5,770.00

Types of cost drivers

1) Transactional cost driver
2) Duration cost driver
3) Budget cost driver

Sometimes it is not clear which cost driver should be used. Although there is no mathematical formula for the choice of cost driver, it is expected that cost drivers meet the following criteria if possible:

- A cost driver has *a cause-and-effect relationship* with a cost object. In other words, an increase in the volume of a cost driver leads to an increase in costs of a cost object.
- The volume of a cost driver can be *tracked and measured in an objective way*. For example, although a factor such as the quality of inspection seems to have a cause-and-effect relationship with costs of inspection, it is not qualified as a cost driver because it is hard to measure and track the quality of inspection in an objective way.
- The volume of a cost driver should be *measured periodically in an economically feasible way*. Even though a cost driver meets the first two requirements, it is not a good cost driver if a significant amount of time is needed to measure the volume of the cost driver. In this case, you may want to use another cost driver, one which reduces the efforts required to track the volume.

Implementation Tip

You need to limit the number of cost drivers in developing your ABC systems because you need to track the volume of cost drivers consumed by cost objects every month. You might find that multiple activities share cost drivers, so that efforts to track the volume of cost drivers are reduced.

2.5   Chapter summary

In this chapter, we discussed ABC compared with a traditional overhead costing method. The traditional costing method assumes one-stage costing in which products or services directly consume resources. It allocates resource costs to cost objects using volume-based allocation, such as direct labor costs. However, the traditional costing method has limitations for two reasons:

1) it is likely to have cost distortion
2) it does not provide cost information on activities (the lack of process view).

On the other hand, ABC uses two-stage costing in which resource costs are assigned to activities and activity costs are assigned to cost objects. The literature shows that ABC has two distinct advantages compared to the traditional costing method: (1) ABC allocates overhead costs to cost objects accurately, and (2) ABC enables users to identify non-value-adding activities for process improvement.

In Chapter 3 we will discuss the first case of ABC application: ABC applied to managing project overhead costs.

References

Ballard, G., Kim, Y., Min, L., and Jang, J. (2007). *Lean Implementation at a Project Level*, Construction Industry Institute, Research Report 234, Austin, TX.

Brimson, J. (1998). "Featuring costing: beyond ABC," *Journal of Cost Management*, Jan/Feb, 6–12.

Cokins, G. (1996). *Activity-based Cost Management Making it Work: a Manager's Guide to Implementing and Sustaining an Effective ABC system*, Irwin Professional Pub., Burr Ridge, IL.

Cooper, R. (1989). "The rise of activity based costing," *Journal of Cost Management*, Spring, 38–49.

Cooper, R. and Kaplan, R. (1992). "Activity-based systems: Measuring the costs of resource usage," *Accounting Horizons* (September) 1–13.

Horngren, C., Foster, G., and Datar, S. (1999). *Cost Accounting*, 10th edn, Prentice Hall, Upper Saddle River, NJ.

Johnson, T. (1992). *Relevance Regained: From Top-Down Control to Bottom-Up Empowerment*, Free Press, New York, NY.

Johnson, T. and Kaplan, R. (1987). *Relevance Lost: Rise and Fall of Management Accounting*, Harvard Business School Press, Boston, MA.

Kaplan, R. and Cooper, R. (1997). *Cost and Effects: Using Integrated Cost Systems to Drive Profitability and Performance*, Harvard Business School Press, Boston, MA.

Kim, Y. (2002). *The Implications of a New Production System for Project Cost Control*, PhD thesis, Dept. of Civil Engineering, University of California at Berkeley, Berkeley, CA, December, 2002.

Koskela, L. (1992). *Application of the New Production Philosophy in the Construction Industry*, Technical report No. 72, Center for Integrated Facilities Engineering, Department of Civil Engineering, Stanford University, CA.

Miller, J. (1992). "Designing and implementing a new cost management system," *Journal of Cost Management*, Winter, 41–53.

Miller, J. (1996). *Implementing Activity-Based Management in Daily Operations*, John Wiley and Sons, New York, NY.

Raffish, N. and Turney, P. (1991). "Glossary of activity-based management," *Journal of Cost Management*, Fall, 53–63.

3 Managing Overhead Costs in Construction Projects

Activity Based Costing for Construction Companies, First Edition. Yong-Woo Kim.
© 2017 John Wiley & Sons Ltd. Published 2017 by John Wiley & Sons Ltd.

One characteristic of current construction projects is the increased use of specialty contractors. As a result, projects grow more complicated and fragmented; general contractors are spending more time and effort on coordinating their subcontractors. However, subcontractors working on fixed-priced contracts may reduce the number of management staff to lower their overhead costs, even at the risk of increasing their total costs through a lack of management. Moreover, new management principles and technologies such as Lean construction, BIM (Building Information Modeling), and sustainable design and construction have caused general or project overhead costs to increase.

In such environments, general contractors' overhead costs have increased relative to direct costs (Kim and Ballard, 2001; Kim and Ballard, 2005). However, the construction industry has focused primarily on project direct costs, as opposed to overhead costs. As for industry practitioners, overhead costs have generally been controlled mainly for tax and financial purposes, not for managerial purposes (Coombs and Palmer, 1989). A method of analyzing project overhead costs is much needed. In this chapter, we will discuss what managing project overhead costs means and how activity-based costing (ABC) can be applied to managing project overhead costs.

3.1   Project overhead costs as profit points

Suppose that contracts between a general contractor and specialty contractors are fixed-price or lump-sum contracts. Assume that there is a general contractor using 100% outsourcing in performing a project. Given this situation the profit can be calculated as follows:

[Equation 3.1]

$$
\begin{aligned}
\text{Profit} = &(1)\,\text{Total contract amount}\,(\text{Revenue}) - \\
&(2)\,\text{Total of subcontract amount}\,(\text{Direct cost}) - \\
&(3)\,\text{Project management costs} - \\
&(4)\,\text{Sustaining costs}
\end{aligned}
$$

Project management costs refer to costs associated with managing projects. Resources consumed by activities associated

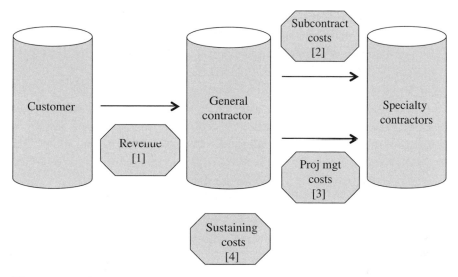

Figure 3.1   Cost flow in a construction project.

with project management such as scheduling, cost control, quality control, and safety management are included in this category. The term "sustaining costs" refers to all overhead costs except project management costs. Examples in the category of sustaining costs are project office rent and office supplies. Project management costs in Equation 3.1 depend largely on managing specialty contractors, each of whom is in charge of one or a couple of work divisions.

Figure 3.1 shows the revenue and cost flow of a construction project from the perspective of a general contractor using 100% outsourcing. The flow of profit becomes visible when costs are revealed where revenues are created. In a construction project, apart from additional change orders, revenue is fixed when a fixed-price contract is made between a company (general contractor) and a client (Kim and Ballard, 2005). In most cases, revenue from a project can be allocated or distributed to different work divisions or different facilities. Outsourcing costs (i.e., subcontract costs), which are part of project direct costs, are determined when a fixed-priced contract is made between a general contractor and a specialty contractor. Given that revenue and direct costs are determined by contracts, a general contractor's profits are heavily dependent on project overhead costs, which include (3) management costs and (4) sustaining costs. In summary, managing project overhead costs is critical to a contractor's profits.

The areas where project overhead costs occur are called profit points (Kim and Ballard, 2005). However, the flow of cost and profit is not visible to general contractors because overhead cost data comes from a single overhead cost pool into which a traditional costing method combines all project overhead costs. The practice of consolidating all project overhead costs into a single cost pool conceals the process waste that overhead resources generate (Johnson and Broms, 2000). Profit points are also points where a general contractor and specialty contractors interface (Kim and Ballard, 2005). In this regard, revealing activities and their costs on profit points would help a contractor improve its collaboration with subcontractors.

3.2   Implementing ABC to manage project overhead costs

In Chapter 2, we discussed general ABC methodology and listed the major steps in the implementation of ABC. In this chapter, we will examine a case where ABC was applied successfully to managing construction project overhead costs. I hope that it will give you guidance when you develop your own ABC system for managing your project overhead costs.

3.3   Case study: xx Commercial Complex[1]

This case concerns a commercial complex which consisted of three buildings. The general contractor embarked on an ABC pilot study in June 2008. Management wanted to apply ABC to their projects to answer the following questions:

- Usage of project overhead costs: How much overhead resources would be needed to manage a project?
- Evaluation of subcontractors: How can we use overhead cost information to evaluate each subcontractor's performance?

First, the company wanted to understand the level of resources consumed in each work division and for each project management function. The information is needed to accurately estimate

---

[1] The numbers in the case study have been changed for confidentiality.

project overhead costs in future projects, assuming that the ratio of project overhead costs to total project costs in similar projects does not vary significantly.

Second, the contractor assumed that additional overhead resources (management staff) would be needed to manage its subcontractors if they did not bring good quality workforces to the project. Project managers had complained that sometimes subcontractors did not use quality manpower, including management staff such as project engineers, in order to reduce their costs because the company (a general contractor) selected subcontractors on the basis of the lowest bids. Failure to bring quality manpower usually leads to a general contractor's overhead costs being increased to manage those subcontractors. However, the current costing method does not provide a general contractor with any help in effectively evaluating its subcontractors' quality aspects. In this case study, management expected that the ABC system would allow for quantitative analysis of subcontractors' performance.

3.3.1   *Developing an activity-based costing charter*

First, the task force was made up of three individuals: an outside facilitator, an employee from the project management team, and an employee from the construction team (i.e., the field team). The team had an executive sponsor who was an operations manager of the company. The team had a kick-off meeting where a team charter was developed (Figure 3.2). Some components included in the charter were:

- *Scope*: The team decided to include only human resources because human resources are the only resource relevant to two goals pre-set by management. Note that it is nevertheless feasible to include other resources such as facility and material costs.
- *Objectives*: As mentioned, the management had set two goals: (1) monitoring where and why project overhead costs are spent and (2) evaluating their subcontractors. The team developed five concrete objectives to achieve these two goals:
  1) to identify major activities
  2) to estimate activity costs
  3) to calculate cost driver rates

ACTIVITY-BASED COSTING CHARTER

FACILITIES

1. Human resource project overhead costs

2. NOT INCLUDED: facility, utilities, materials

OPERATIONS MANAGERS	
Executive sponsor:	xxxx (operation manager)
1. Facilitator:	xxxx
2. PM team:	xxxx
3. Construction team:	xxxx

OBJECTIVES

1. Identify major activities

2. Identify activity costs (esp. safety costs)

3. Identify costs to manage each work division

4. Evaluate subcontractors

LOGISTICS	
Regular meeting	5:00 PM-6:00 PM on M;W;F
Workshop	2 hrs, all project members are expected to participate.
Interim report	by 5:00 PM July x, 20xx
Final report	by 5:00 PM July x, 20xx

GROUND RULES

1. Deliver each assignment to the facilitator at least 3 hrs before each meeting

AGREEMENT

Team members' signature

APPROVED BY

Executive sponsor:

Project manager:

Superintendent:

Figure 3.2   Team charter example.

    4) to evaluate management areas to identify areas and activities for improvement

    5) to evaluate subcontractors.

- *Logistics*: The total timeframe for developing the ABC system was one month. In addition to regular team meetings, the team decided to set up a workshop where every project member could participate before the team began to collect data.

- *Ground rules*: The team decided to develop a simple rule: everyone does his or her assignment and distributes it to the team members prior to each meeting. The team needed this rule to make regular meetings efficient, allowing them to focus on problem solving, instead of information sharing.
- *Agreement*: In addition to the agreement requiring each team member's signature, the team added a section for the executive sponsor, the project manager, and the superintendent to sign (this was placed approve the charter). This type of approval process was necessary, because the data collection task needed support from both project management and the construction field team.

You can use variations of the charter shown in Figure 3.2. For example, you can include details of data collection methods and/or data analysis methods if you feel it necessary. If you prefer to create your own charter, it is recommended that you include at least the five elements described above.

Implementation Tip

In developing your ABC system for managing project overhead costs, your task force needs to include at least one member from your construction field team, such as a field engineer. In addition, you need to ask for help from a superintendent when your ABC project begins.

In some cases, a construction field team may not support the data collection process. Failure to get support from a field team frequently leads to the failure of an ABC system because the field team's input is critical to developing the ABC system.

3.3.2 Workshop

The workshop had two main functions:

1) to provide information about the ABC system and the data collection process; and
2) to ask for help and support from each individual.

A key criterion in the workshop setting was that all the main personnel (i.e., heads of each department) attended. It was recognized from the beginning that a lack of understanding could lead to confrontation. One of the common misunderstandings is that the collection of data on time spent on activities is intended to track the efficiency of each department or of personnel. Such misunderstanding may lead to uncooperative behavior in relation to data collection, a task which usually involves surveys or interviews. In the two-hour workshop, the task force focused on (1) the purpose of the ABC system and (2) the level of activity detail that the task force expected.

3.3.3   Defining cost objects

The cost objects in your ABC system are determined by finalizing the system objectives and the system scope in your charter. The team defined three types of cost objects to achieve its goals: (1) management areas, (2) buildings, and (3) work divisions.

3.3.3.1   Management areas

The first type of cost object is a management area such as safety control. With cost information on each management area, you can find how much project overhead resources are spent in managing each management area. Cost information on management areas can provide insight into areas where you need to improve efficiency when compared to other projects. The list of cost objects in the management area is as follows:

1) Cost management
2) Quality management
3) Time management
4) Safety management
5) Procurement management
6) General management
7) Design management (or scope management).

3.3.3.2   Buildings

The project consisted of three buildings. The team discussed whether these buildings should be cost objects. Although it was not required to identify overhead costs for each building in this project, the information would be needed for future reference (e.g., if each building was to be sold or leased separately). Therefore, the team decided to include the buildings as cost objects to allow for future reference. This list of cost objects is as follows:

1) Building #1
2) Building #2
3) Building #3.

3.3.3.3 Work divisions

The last type of cost object is a work division. Although the CSI (Construction Specification Institute) code lists 50 work divisions in its MasterFormat (Johnson, 2004), the team used only six major work divisions. You can develop more work divisions based on your needs. The list of cost objects was developed as follows:

1) Earthwork
2) Structure
3) Finishes
4) Cladding
5) Mechanical/plumbing
6) Electrical.

Cost information on multiple cost objects can give insight into overhead costs and their relationship with specialty contractors. This information is important because the activities which consume your overhead resources are the hub of your business for a project. In contrast, other accounting systems put all overhead cost information into cost accounts which combine the profit points.

3.3.4 Identifying activities

One of the difficult tasks in identifying activities is determining the level of detail. When you ask people "what activities do you perform?" the answers may vary from activities that are too detailed (e.g., receiving phone calls or having a specific meeting) to activities that are too abstract (e.g., quality control). As mentioned in Chapter 2, an activity is a concept larger than a daily task, but smaller than a process or a function. Why is it a problem if activities are too detailed? In practical terms, the volume of cost drivers needs to be tracked periodically (e.g., every month or every week). If the activities are too detailed, it is really hard to track the volume of cost drivers. However, in contrast to overly detailed activities, activities that are overly aggregated do not provide insight into process.

The two-hour workshop helped project members understand the level of detail required for defining activities for the ABC system. Following the workshop, the team had several rounds of interviews with project members to identify activities. After several discussions among the task force members, the team finally identified 20 activities. The activities were then classified into seven activity centers, corresponding to management areas, one of the cost objects already defined (see Table 3.1). It is recommended that you develop this type of table showing a hierarchical structure.

You may be tempted to have more detailed activities in your list. For example, an activity of "inspection" may have sub-activities such as requesting inspection and retrieving updated drawings and specifications (see Figure 3.3). The detailed sub-activities are suitable for process improvements, but not for an ABC system that needs to be updated regularly.

3.3.5 *Assigning resource costs to activities*

One of the first things your team has to decide is the method of assigning resource costs to activities. There were two ways of assigning human resource costs to activities. They are:

1) tracking the time each person spends on each activity, and
2) the time–effort percentage (time–effort %) method, whereby the relevant percentage of each person's time is allocated to each activity.

The first method (i.e., tracking each person's time) can be more accurate but also more time-consuming. This method also places the burden on employees (or project team members) because they have to track their time not only for developing the ABC system but also for updating the system. For this reason, the team chose to use the second method, the time–effort % method.

Table 3.1 Activity centers and activities.

Activity center	Activity ID	Activity
Cost management	1	Cost reporting (head office)
	2	Requesting payments
Quality management	3	Supervising field activities
	4	Inspection
	5	Handling NCR[1] report
	6	Testing (concrete)
Time management	7	Developing/updating project scheduling
	8	Monthly/weekly production scheduling
Safety management	9	Safety training
	10	Safety report
Procurement management	11	Requesting material purchase
	12	Monthly material management reporting
	13	Coordinating delivery schedule with suppliers
	14	Handling deliveries
	15	Preparing submittals
General management	16	General correspondence
	17	Manpower hiring/management
Design/scope management	18	Reviewing drawings
	19	Issuing RFI[2]
	20	Change order issue

[1] NCR (Non-conformance report): a report issued when a product or work does not comply with drawing and specification.
[2] RFI (Request for information): a written memo that request clarification or additional information.

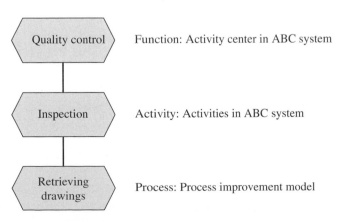

Quality control — Function: Activity center in ABC system

Inspection — Activity: Activities in ABC system

Retrieving drawings — Process: Process improvement model

Figure 3.3 Example of activity granularity.

Below are the step-by-step procedures that the team went through to assign resource costs to activities when the time–effort % method was employed.

1) The task force team listed activities and employees' names as shown in Table 3.2.
2) The task force held interviews with the project team members (i.e., those listed in Table 3.2). Note that interviewees sometimes feel uncomfortable about tracking their time. Therefore, it is necessary for you to make certain that the information will not be used to evaluate interviewees' performance or productivity. Without this assurance, people tend to manipulate the numbers (i.e., time spent on activities) to make them look productive or valuable.

   It is true that there may be activities on which employees spend time but which are not in the activity list. This occurs because only major activities are captured in the activity list in ABC systems. Therefore, we do not expect accurate allocation; instead, the result is a reasonable allocation of resource costs to activities. Table 3.3 shows the result of the assignment of employees' time to activities using the time–effort % method.
3) The next step is to calculate activities costs using unit resource cost. In this case study, the only unit resource cost is the employees' salaries. Usually you can get employee salary information from a company's human resources department or accounts department.

   The monthly activities' costs can be calculated as follows:

$$\sum_{i=1}^{n} r_i \times ms_i$$

where,
i = the resource number (or the employee number)
ms_i = the monthly salary of resource i
r_i = time–effort % of resource i on the activity

For example, the activity cost of "inspection" can be calculated as:
5% × \$11,000 + 20% × \$8,200 + 15% × 6,800 + 27% × 9,300 = \$5,721
Table 3.4 shows the result of the monthly activity costs.

Table 3.2 Template time–effort % table.

List of Employees	List of Activities					
	Cost reporting	Requesting payments	...	RFI issue	Change order issue	Total Percentage
Blackburn						100%
Zou						100%
Carey						100%
Coombs						100%
Bailey						100%
Cox						100%
McDonough						100%
Carroll						100%
Kollman						100%

Table 3.3 Time–effort % table.

	Blackburn	Zou	Carey	Coombs	Bailey	Cox	McDonough	Carroll	Kollman
Cost reporting	25%	30%							
Requesting payments	15%	30%							
Supervising field activities				25%	20%	28%	26%	15%	
Inspection					5%	20%	15%	27%	
Handling NCR report	3%				7%	10%	7%	24%	
Testing (concrete)								30%	
Project scheduling	10%	5%			10%	7%	8%		
Production planning		5%			13%	20%	22%		
Safety training				25%	5%				10%
Safety reporting				45%					
Requesting material purchase			10%		5%				
Monthly material management report	8%		15%						
Coordinating with suppliers	3%		30%						
Handling deliveries			30%		2%	5%	7%		
Preparing submittals			15%			5%	5%		
General correspondences	11%	10%		5%				4%	52%
Manpower hiring/management	10%				3%				28%
Drawing review	5%				10%				
RFI issue	10%				10%	5%	10%		10%
Issuing change order	10%				10%				
Percentage	100%	100%	100%	100%	100%	100%	100%	100%	100%

3.3.6   *Assigning activity costs to cost objects*

The next step is to assign activity costs to cost objects. However, you cannot assign activity costs to cost objects without a properly defined cost driver. Therefore, you need to choose an appropriate cost driver for each activity.

3.3.6.1   Identifying a cost driver for each activity

The team held two meetings to set up cost drivers. The team used three criteria in selecting cost drivers: each driver needed:

- to have a cause-and-effect relationship;
- to be capable of being measured in an objective way;
- to be capable of being measured in an economically feasible way.

The following section addresses how the team chose the cost driver for each activity. The complete set of cost drivers is shown in Table 3.5.

Activity Center #1: Cost management

There are only two activities in the activity center of cost management. Although one member proposed using the number of budget items (or lines) to be used a cost driver, the team rejected it because counting the number of budget items in every cost report seemed overly time-consuming. The team decided to use a budgetary cost driver; the activity costs would be assigned to cost objects based on the ratio of total budget to the budget of each cost object.

Activity Center #2: Quality management

In terms of the "supervising field operations" activity, the team first thought that supervising time might be a good cost driver because (1) supervising time has a cause-and-effect relationship with the activity cost of "supervising," and (2) it can be measured in an objective way. However, tracking supervising time for each cost object (i.e., work divisions and buildings) seemed time-consuming and painful.

As an alternative, the team agreed to use a transactional cost driver for activities associated with inspection because it met all three criteria.

Activity Center #3: Time management

Developing and updating project schedules and production plans (i.e., short-term schedules) are activities for which the cost drivers are hard to define. Both a budgetary cost driver and a

Table 3.4 Monthly activity costs.

	Cost reporting	Requesting payments	Supervising field activities	Inspection	Handling NCR reports	Testing (concrete)	Project scheduling	Production Planning	Safety Training	Safety reporting
Blackburn	25% $3,000.00	15% $1,800.00	$-	$-	3% $360.00		10% $1,200.00	$-		
Zou	30% $2,550.00	30% $2,550.00	$-	$-	$-					$-
Carey			$-	$-	$-		5% $425.00	5% $425.00		$-
Coombs			25% $2,300.00						25% $2,300.00	45% $4,140.00
Bailey			20% $2,200.00	5% $550.00	7% $770.00	$-	10% $1,100.00	13% $1,430.00	5% $550.00	$-
Cox			28% $2,296.00	20% $1,640.00	10% $820.00	$-	7% $574.00	20% $1,640.00	$-	$-
McDonough			26% $1,768.00	15% $1,020.00	7% $476.00	$-	8% $544.00	22% $1,496.00	$-	$-
Carroll			15% $1,395.00	27% $2,511.00	24% $2,232.00	30% $2,790.00	$-	$-	$-	$-
Kollman									10% $360.00	$-
Activity Costs	$5,550.00	$4,350.00	$9,959.00	$5,721.00	$4,658.00	$2,790.00	$3,843.00	$4,991.00	$3,210.00	$4,140.00

Requesting Material Purchase	Monthly Material Management Report	Coordinating with Suppliers	Handling Deliveries	Preparing Submittals	General Correspondences	Manpower Hiring/Management	Drawing Review	RFI Issue	Issuing change order	Percentage
$-	8% $960.00	3% $360.00	$-	$-	11% $1,320.00	10% $1,200.00	5% $600.00	$-	10% $1,200.00	100% $12,000.00
$-	$-	$-	$-	$-	10% $850.00	$-	$-	10% $850.00	10% $850.00	100% $8,500.00
10% $780.00	15% $1,170.00	30% $2,340.00	30% $2,340.00	15% $1,170.00	$-	$-	$-	$-	$-	100% $7,800.00
$-	$-	$-	$-	$-	5% $460.00	$-	$-	$-	$-	100% $9,200.00
5% $550.00	$-	$-	2% $220.00	$-	$-	3% $330.00	10% $1,100.00	10% $1,100.00	10% $1,100.00	100% $11,000.00
$-	$-	$-	5% $410.00	5% $410.00	$-	$-	$-	5% $410.00	$-	100% $8,200.00
$-	$-	$-	7% $476.00	5% $340.00	$-	$-	$-	10% $680.00	$-	100% $6,800.00
$-	$-	$-	$-	$-	4% $372.00	$-	$-	$-	$-	100% $9,300.00
$-	$-	$-	$-	$-	52% $1,872.00	28% $1,008.00	$-	10% $360.00	$-	100% $3,600.00
$1,330.00	$2,130.00	$2,700.00	$3,446.00	$1,920.00	$4,874.00	$2,538.00	$1,700.00	$3,400.00	$3,150.00	$76,400.00

Table 3.5   List of cost drivers.

Activity Center	Activity	Cost driver	Cost driver type
Cost management	Cost reporting	Budget (costs ratio)	Budgetary
	Requesting payments	Budget (costs ratio)	Budgetary
Quality management	Supervising field activities	Budget (costs ratio)	Budgetary
	Inspection	The number of inspections	Transactional
	Handling NCR report	The number of NCRs issued	Transactional
	Testing (concrete)	The number of samples	Transactional
Time management	Project scheduling	The number of milestones	Transactional
	Production planning	The number of milestones	Transactional
Safety management	Safety training	Budget	Budgetary
	Safety reporting	The number of violations	Transactional
Procurement management	Requesting material purchase	The number of requests	Transactional
	Monthly material management report	Budget (costs ratio)	Budgetary
	Coordinating with suppliers	The number of deliveries	Transactional
	Handling deliveries	The number of deliveries	Transactional
	Preparing submittals	The number of submittals	Transactional
General management	General correspondence	Budget (costs ratio)	Budgetary
	Manpower hiring/ management	Budget (costs ratio)	Budgetary
Design/scope management	Drawing review	The number of drawing sheets	Transactional
	RFI issue	The number of RFIs	Transactional
	Change order issue	The number of change orders	Transactional

transactional cost driver were discussed. The team first rejected a budgetary cost driver because the costs of scheduling are not proportional to budget; instead, they have a correlation with the number of activities involved in scheduling. However, the number of activities seemed too detailed and not practical in measuring the volume of the cost driver. Finally, the team chose the number of milestones on which short-term schedules were developed and updated.

Activity Center #4: Safety management

The safety management activity center had two activities: safety training and safety reporting. In the meetings, the team easily agreed that the effort of developing a safety report correlated to the number of safety violations because a significant portion of each safety report addressed the causes of and actions on each safety violation case. Therefore, the team chose the number of safety violations as a cost driver for the activity of safety reporting.

While choosing a cost driver for the "safety reporting" activity was straightforward, identifying a cost driver for the "safety training" activity was difficult because workers from different work divisions are co-mingled in a single safety training session. In addition, it was hard to distinguish training hours or sessions by reference to a particular building. For these reasons, the team chose to use a budgetary cost driver (i.e., labor cost ratio) as a cost driver for the "safety training" activity.

Activity Center #5: Procurement

The team agreed to use a transactional cost driver for the procurement activities except for the "monthly material management reporting" activity. It was hard to identify a transactional cost driver for the monthly material management reporting activity. In particular, the team had a hard time identifying a factor that had a cause-and-effect relationship with the costs of this activity. Although consensus was not reached, the team chose to use a budgetary cost driver (i.e., material costs ratio) for the "monthly material management reporting" activity.

Activity Center #7: General management

Due to the difficulty of identifying a transactional cost driver, the team decided to use a budgetary cost driver for activities in the general management activity center. The total budget ratio of cost objects is a cost driver for the general management activities: (1) general correspondence and (2) resource hiring and management.

Activity Center #8: Design and scope management
The team decided to use a transactional cost driver for all three activities in this activity center. The number of drawing sheets, the number of RFIs (requests for information), and the number of change orders were used as cost drivers. All three cost drivers were easy to measure and track.

3.3.6.2   Calculating a unit rate of activity costs

The unit rate of an activity cost (also known as a unit activity cost) can be defined as follows:
[Equation 3.2]

$$Unit\ rate\ of\ activity\ costs = \frac{Activity\ cost}{Total\ volume\ of\ a\ cost\ driver}$$

The unit rate of an activity cost was used as the cost-allocation base. In this step, you need to measure the total volume of each cost driver in each period. In this project, the task force measured the volume of cost drivers for the previous one month. The unit rate of activity costs is presented in Table 3.6.

3.3.6.3   Assigning activity costs to cost objects using a unit rate of activity costs

Using unit rates of activity costs (i.e., the allocation base), you can easily calculate the costs of cost objects using Equation 3.2. Tables 3.7 and 3.8 show how the costs of cost objects are calculated. Assigning activity costs to the activity centers does not need cost driver information because the sum of activity costs relevant to each activity center equals the costs of each activity center.

3.4   Using ABC data for managerial purposes

Knowing how to use ABC data is as important as knowing how to develop an ABC system. The task force developed a strategic plan for using ABC data to achieve the goals set up in the charter. There were four goals established in this ABC project:

1)   To identify major activities
2)   To identify activity costs
3)   To evaluate management areas to identify areas and activities for improvement
4)   To evaluate subcontractors.

Table 3.6 Unit rate of activity costs.

Activity ID	Activity	Cost driver	Total activity cost	Vol. of cost drivers	Unit activity cost
1	Cost reporting	Budget	$5,550.00	22	$252.27
2	Requesting payments	Budget	$4,350.00	22	$197.73
3	Supervising field activities	Budget	$9,959.00	22	$452.68
4	Inspection	# of inspections	$5,721.00	64	$89.39
5	Handling NCR reports	# of NCRs issued	$4,658.00	6	$776.33
6	Testing (concrete)	# of samples	$2,790.00	32	$87.19
7	Project scheduling	# of milestones	$3,843.00	15	$256.20
8	Production planning	# of milestones	$4,991.00	15	$332.73
9	Safety training	Budget	$3,210.00	22	$145.91
10	Safety report	# of violations	$4,140.00	7	$591.43
11	Requesting material purchase	# of requests	$1,330.00	12	$110.83
12	Monthly material management report	Budget	$2,130.00	22	$96.82
13	Coordinating with suppliers	# of deliveries	$2,700.00	27	$100.00
14	Handling deliveries	# of deliveries	$3,446.00	27	$127.63
15	Preparing submittals	# of submittals	$1,920.00	18	$106.67
16	General correspondence	Budget	$4,874.00	22	$221.55
17	Manpower hiring/management	# of direct labor	$2,538.00	22	$115.36
18	Drawing review	# of drawing sheets	$1,700.00	14	$121.43
19	RFI issue	# of RFIs	$3,400.00	12	$283.33
20	Change order issue	# of change orders	$3,150.00	3	$1,050.00
	Total		$76,400.00		

Table 3.7 Overhead costs of the cost object (buildings).

Activity ID	Activity	Unit Rate	Building #1		Building #2		Building #3		Total vol., cost driver	Activity costs
			Volume, cost driver	Activity costs	Volume, cost driver	Activity costs	Volume, cost driver	Activity costs		
1	Cost reporting	$252.27	6.3	$1,589.32	7.3	$1,841.59	8.4	$2,119.09	22	$5,550.00
2	Requesting payments	$197.73	6.3	$1,245.68	7.3	$1,443.41	8.4	$1,660.91	22	$4,350.00
3	Supervising field activities	$452.68	6.3	$2,851.90	7.3	$3,304.58	8.4	$3,802.53	22	$9,959.00
4	Inspection	$89.39	21	$1,877.20	18	$1,609.03	25	$2,234.77	64	$5,721.00
5	Handling NCR reports	$776.33	1	$776.33	3	$2,329.00	2	$1,552.67	6	$4,658.00
6	Testing (concrete)	$87.19	10	$871.88	10	$871.88	12	$1,046.25	32	$2,790.00
7	Project scheduling	$256.20	5	$1,281.00	5	$1,281.00	5	$1,281.00	15	$3,843.00
8	Production planning	$332.73	5	$1,663.67	5	$1,663.67	5	$1,663.67	15	$4,991.00
9	Safety training	$145.91	6.3	$919.23	7.3	$1,065.14	8.4	$1,225.64	22	$3,210.00
10	Safety reporting	$591.43	3	$1,774.29	2	$1,182.86	2	$1,182.86	7	$4,140.00
11	Requesting material purchase	$110.83	4	$443.33	3	$332.50	5	$554.17	12	$1,330.00
12	Monthly material management report	$96.82	6.3	$609.95	7.3	$706.77	8.4	$813.27	22	$2,130.00

#	Description								
13	Coordinating with suppliers	$100.00	8	$800.00	10	$1,000.00	9	$900.00	27	$2,700.00
14	Handling deliveries	$127.63	8	$1,021.04	10	$1,276.30	9	$1,148.67	27	$3,446.00
15	Preparing submittals	$106.67	5	$533.33	6	$640.00	7	$746.67	18	$1,920.00
16	General correspondence	$221.55	6.3	$1,395.74	7.3	$1,617.28	8.4	$1,860.98	22	$4,874.00
17	Manpower hiring/ management	$115.36	6.3	$726.79	7.3	$842.15	8.4	$969.05	22	$2,538.00
18	Drawing review	$121.43	4	$485.71	5	$607.14	5	$607.14	14	$1,700.00
19	RFI issue	$283.33	3	$850.00	6	$1,700.00	3	$850.00	12	$3,400.00
20	Issuing change orders	$1,050.00	1	$1,050.00	1	$1,050.00	1	$1,050.00	3	$3,150.00
	TOTAL			$22,766.39		$26,364.29		$27,269.32		$76,400.00

Table 3.8 Overhead costs of the cost object (work divisions).

Activity	Earthwork		Structure		Finish		Cladding		Mechanical		Electrical		Total	
	Volume, cost Driver	Activity costs	Volume, cost Driver	Activity costs	Volume, cost Driver	Activity costs	Volume, cost Driver	Activity costs	Volume, cost Driver	Activity costs	Volume, cost Driver	Activity Costs	Volume, Cost Driver	Activity Costs
Cost reporting	3.4	$857.73	6.9	$1,740.68	2.5	$630.68	3.6	$908.18	3.2	$807.27	2.4	$605.45	22	$5,550.00
Requesting payments	3.4	$672.27	6.9	$1,364.32	2.5	$494.32	3.6	$711.82	3.2	$632.73	2.4	$474.55	22	$4,350.00
Supervising field activities	3.4	$1,539.12	6.9	$3,123.50	2.5	$1,131.70	3.6	$1,629.65	3.2	$1,448.58	2.4	$1,086.44	22	$9,959.00
Inspection		$-	31	$2,771.11		$-	13	$1,162.08	14	$1,251.47	6	$536.34	64	$5,721.00
Handling NCR report		$-	4	$3,105.33		$-		$-	2	$1,552.67		$-	6	$4,658.00
Testing (concrete)		$-	32	$2,790.00		$-		$-		$-		$-	32	$2,790.00
Project scheduling		$-	5	$1,281.00	2	$512.40	3	$768.60	3	$768.60	2	$512.40	15	$3,843.00
Production planning		$-	5	$1,663.67	2	$665.47	3	$998.20	3	$998.20	2	$665.47	15	$4,991.00
Safety training	3.4	$496.09	6.9	$1,006.77	2.5	$364.77	3.6	$525.27	3.2	$466.91	2.4	$350.18	22	$3,210.00
Safety reporting		$-	3	$1,774.29		$-		$-	3	$1,774.29	1	$591.43	7	$4,140.00

Activity	Qty	Cost	Qty	Cost	Qty	Cost	Qty	Cost	Qty	Cost	Qty	Cost	Count	Total
Requesting material purchase		$-	5	$554.17	2	$221.67	5	$554.17		$-		$-	12	$1,330.00
Monthly material mgt report	3.4	$329.18	6.9	$668.05	2.5	$242.05	3.6	$348.55	3.2	$309.82	2.4	$232.36	22	$2,130.00
Coordinating with suppliers		$-	10	$1,000.00	3	$300.00	7	$700.00	5	$500.00	2	$200.00	27	$2,700.00
Handling deliveries		$-	10	$1,276.30	3	$382.89	7	$893.41	5	$638.15	2	$255.26	27	$3,446.00
Preparing submittals		$-		$-	7	$746.67	2	$213.33	5	$533.33	4	$426.67	18	$1,920.00
General correspondence	3.4	$753.25	6.9	$1,528.66	2.5	$553.86	3.6	$797.56	3.2	$708.95	2.4	$531.71	22	$4,874.00
Manpower hiring/ management	3.4	$392.24	6.9	$796.01	2.5	$288.41	3.6	$415.31	3.2	$369.16	2.4	$276.87	22	$2,538.00
Drawing review		$-	6	$728.57	1	$121.43	2	$242.86	3	$364.29	2	$242.86	14	$1,700.00
RFI issue		$-	4	$1,133.33		$-	2	$566.67	4	$1,133.33	2	$566.67	12	$3,400.00
Issuing change orders		$-	2	$2,100.00		$-		$-	1	$1,050.00		$-	3	$3,150.00
TOTAL		$5,039.88		$30,405.76		$6,656.31		$11,435.65		$15,307.74		$7,554.65		$76,400.00

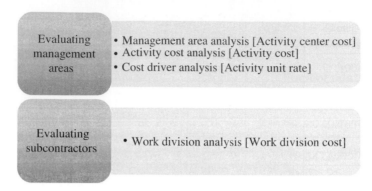

Figure 3.4   Strategic plan for the use of ABC data.

Since the first two goals were achieved as the ABC system was developed, the strategic plan focused on the third and fourth goals. At the team meetings, the team developed a strategic plan which showed how ABC data could be used to achieve these last two goals (Figure 3.4).

3.4.1   Evaluating management areas with activity analysis

The first objective was to evaluate business processes. To this end, the team decided to implement the following analyses:

- Management area analysis
- Activity costs analysis
- Cost driver analysis.

3.4.1.1   Management area analysis

The analysis of management areas used the cost data for each activity center, and identified major processes or functions along the project's value chain that provided support to direct construction works. It provided management with a process view of major processes. Therefore, the analysis data, when compared to data on other projects, enabled management to identify management areas for improvement.

Table 3.9 shows the result of management area analysis (the management areas are ranked in order of overhead costs). As shown in Table 3.9, quality management and procurement management were identified as major management areas. The project manager was content with the result because the ABC results confirmed the needs of the recently initiated quality program.

Table 3.9   Management area analysis.

Activity Center	$	%
Quality management	$23,128	30.27%
Procurement management	$11,526	15.09%
Cost management	$9,900	12.96%
Time management	$8,834	11.56%
Design/scope management	$8,250	10.80%
General management	$7,412	9.70%
Safety management	$7,350	9.62%
Total	$76,400	100.00%

Table 3.10   Management area analysis (comparison analysis).

Activity Center	Project in the case study		Company average	
	%	Ranking	%	Ranking
Quality management	30.27%	1	20.15%	2
Procurement management	15.09%	2	22.75%	1
Cost management	12.96%	3	8.97%	6
Time management	11.56%	4	13.75%	4
Design/scope management	10.80%	5	12.64%	5
General management	9.70%	6	6.38%	7
Safety management	9.62%	7	15.36%	3
Total	100.00%		100.00%	

When compared with company average figures, the result also enabled management to identify the attributes of the project. For example, the project spent more overhead resources on quality management and cost management as compared to the company average (Table 3.10).

3.4.1.2 Activity cost analysis

The activity cost analysis is not independent of the management area analysis. The result of the management area analysis illustrated the major (or critical) management areas: quality management and procurement management. The team could investigate detailed activities in major management areas, the costs in those areas being identified as higher than those of other management areas. The results of activity cost analysis on (1)

Table 3.11   Activity cost analysis results.

Illustrative data: quality management		
Activity	**Activity costs**	**Percentage**
Supervise field activities	$9,959	43.06%
Inspection	$5,721	24.74%
Handling NCR reports	$4,658	20.14%
Testing (concrete)	$2,790	12.06%
TOTAL	$23,128	100.00%

Illustrative data: procurement management		
Activity	**Activity costs**	**Percentage**
Request material purchase	$1,330	11.54%
Monthly material management report	$2,130	18.48%
Coordinating delivery schedule	$2,700	23.43%
Handling deliveries	$3,446	29.90%
Prepare submittals	$1,920	16.66%
TOTAL	$11,526	100.00%

quality management and (2) procurement management are presented in Table 3.11. "Supervising field activities" and "inspection" are costly activities in the quality management area; "handling deliveries" and "coordinating delivery schedule" are costly activities in the procurement management area (Table 3.11).

The activity cost analysis results provided management with insight into specific activities to which they needed to pay special attention. For example, the activities of "handling deliveries" and "coordinating delivery schedule" consumed more than half of procurement costs.

3.4.1.3   Cost driver analysis

The costs assigned by each cost driver, or cost driver rates, were calculated in Table 3.12. These rates show productivity or efficiency of resources in implementing activities. They can be also used as benchmarks when comparing resource consumption efficiency to that of similar projects.

For example, project managers could investigate how the "handling deliveries" activity could improve if the average cost driver rate of "handling delivery" was below $100. (Note: the cost driver rate of "handling delivery" is $127.63, Table 3.12).

Table 3.12   Cost driver analysis.

Costs per milestone updated	
Project scheduling	$256.20
Short-term production planning	$332.73
Costs per inspection	
Inspection	$89.39
Costs per NCR	
Handling NCR report	$776.33
Cost per delivery	
Coordinating with suppliers	$100.00
Handling deliveries	$127.63
Cost per drawing sheet	
Reviewing drawings	$121.43
Cost per RFI	
Issuing RFI	$283.33
Cost per change order	
Issuing change order	$1,050.00

Implementation Tip

Do you want to reduce your project overhead costs? There are two ways to reduce project overhead costs so as to efficiently use project overhead resources, using an ABC system.

1) The first method is to increase the efficiency of resource consumption per cost driver, or cost driver rate. To do that, you may use ABC information, especially cost driver information as in Table 3.12. What cost driver(s) should you identify to reduce the cost driver rate? The best way is a benchmark system where you make comparison analysis with other projects. Once an activity is identified, you need to make a process study in which a detailed process map is developed.

2) The second method is to reduce the volume of cost drivers consumed by each cost object. For example, if the number of purchase orders is a cost driver, the person in charge of purchase order requests may try to reduce the number of purchase orders by consolidating purchase order requests as long as this does not impact on the delivery schedule. However, this method can be tricky because reducing the volume of cost drivers sometimes impacts on the process quality.

3.4.2 Evaluating subcontractors

General contractors usually select subcontractors with the lowest bid. Although they evaluate subcontractors at the end of each project, most performance evaluations are based on opinion-based surveys with Likert scale scores. Therefore, the performance evaluation of subcontractors rarely becomes part of bid evaluation for selecting a subcontractor. You can easily imagine a case where a low-bid subcontractor is selected although their past performance is not satisfactory. In this regard, ABC data can provide management with insight into the performance of each subcontractor by investigating how much overhead costs were spent on each subcontractor in the past. Under similar project conditions, the ratio of a subcontract amount to overhead costs can be an indicator in evaluating subcontractors. This evaluation assumes that more competitive and well-performing subcontractors consume less general contractor's resources.

Overhead costs spent on each subcontractor can be estimated using costs per work division as shown in Table 3.13. For example, the overhead costs assigned to a mechanical contractor were $15,390.

Suppose that your subcontractors work on more than one project. You can evaluate your subcontractors' performances by calculating the ratio of total subcontracted amount to total overhead costs for each subcontractor. Figure 3.5 shows how overhead cost information for each project can be allocated to each subcontractor working on multiple projects.

We can define a management burden ratio (MBR) for each subcontractor that represents how much of your project overhead resources a subcontractor consumes:

Table 3.13 Work division cost analysis.

Work division	Overhead costs (1 month)
Structure	$30,522.00
Mechanical	$15,390.00
Cladding	$11,465.00
Electrical	$7,647.00
Finish	$6,695.00
Earthwork	$4,681.00

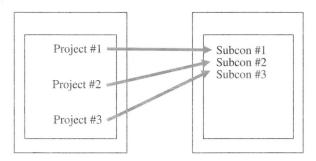

Figure 3.5   Overhead cost allocation to subcontractors.

Table 3.14   Example of MBR comparison.

Subcontractor type: mechanical contractor		
	Total subcontracted amount (2 yrs)	**Total project overhead costs allocated**	**MBR %**
Mechanical contractor A	$32,450,000.00	$1,362,900.00	4.20%
Mechanical contractor B	$28,950,000.00	$781,650.00	2.70%

[Equation 3.3]

$$\text{Management burden ratio \%}$$
$$= \frac{\textit{Project overhead costs allocated to a subcontractor}}{\textit{Total subcontracted amount}}$$
$$= \frac{\sum \textit{Project overhead costs allocated on project i}}{\sum \textit{Subcontracted amount on project i}}$$

Where,
i = Project number

Table 3.14 shows an example where the management burden rates of two mechanical subcontractors are compared based on their subcontracted amounts and the general contractor's project overhead costs allocated to each subcontractor. However, care is needed in using MBR to evaluate subcontractors, because each project context may have an impact on project overhead costs and MBR. Therefore, you should consider the project context when using project overhead costs and MBR in evaluating subcontractors.

3.5 Chapter summary

This chapter covered a case in which ABC was applied to managing project overhead costs. In this chapter, you learned the methodology of implementing ABC in your projects (how ABC can be applied) and how you can use ABC data to effectively leverage your overhead resource consumption as well as to evaluate your subcontractors.

Some of key points covered include:

Ways to get ABC started in your project organization

You need to prepare your project organization prior to launching an ABC system. Two methods are recommended to help your project organization get prepared for ABC:

1) Development of a team charter
2) Workshops.

Flexible cost objects

You can develop various types of cost objects depending on your ABC objectives. Three types of cost objects were used in the case study:

1) Management areas
2) Buildings
3) Work divisions.

In addition, you learned several ways to leverage ABC data to improve your processes as well as to evaluate your subcontractors. ABC also allows contractors to use more accurate cost data when estimating their overhead costs for future projects.

In the next chapter, we will discuss how ABC can be applied to managing your home office overhead costs.

References

Ballard, G. (1994). "The last planner," *Northern California Construction Institute Spring Conference*, Monterey, CA, April 1994.

Coombs, W. and Palmer, W. (1989). *Construction Accounting and Financial Management*, 4th edn, McGraw-Hill, Inc., New York, NY.

Johnson, R. (2004). *Masterformat 2004 Edition: Master List of Numbers and Titles for the Construction Industry.* The Construction Specifications Institute, Alexandria, VA.

Johnson, T. and Broms, A. (2000). *Profit beyond Measure: Extraordinary Results through Attention to Work and People,* Free Press, New York, NY.

Kim, Y. and Ballard, G. (2001). "Activity-based costing and its application to lean construction," *Proceedings of the 9th Annual Conference of the International Group of Lean Construction,* Singapore.

Kim, Y. and Ballard, G. (2005). "Profit point analysis: A tool for general contractors to measure and compare costs of management time expended on different subcontractors," *Canadian Journal of Civil Engineering,* 32(4), 712–718.

4

Managing Your General Overhead Costs

Activity Based Costing for Construction Companies, First Edition. Yong-Woo Kim.
© 2017 John Wiley & Sons Ltd. Published 2017 by John Wiley & Sons Ltd.

In the last chapter, we discussed how the ABC system can be applied to managing project overhead costs. This chapter will discuss general overhead costs or home office overhead costs. First we will discuss what general overhead costs mean and the role of managing general overhead costs before moving on to look at how ABC can be applied to managing general overhead costs.

4.1   General overhead costs

How can we define general overhead costs? Generally speaking, general overhead costs refer to overhead costs incurred at your home office, not on your projects. Strictly speaking, however, this is not true. Suppose that you have an LEED specialist who works at your home office but spends most of his time consulting on LEED accreditation for specific projects. In that case, his wage needs to be charged to the projects he consults on. In other words, his wage is not included in general overhead; instead, his wage is a part of project overhead costs. By definition,[1] general overhead costs are costs not readily chargeable to a particular project (Horngren *et al.*, 1999). They are part of a general contractor's costs, and therefore need to be covered by revenue from projects.

Major items included in a contractor's general overhead costs are as follows:

Office expenses: This category includes any office rent or ownership expenses, office supplies, furniture, utilities (phone, electricity, and water bills), and various taxes.
Vehicle expenses: This category includes car and truck expenses that office and general management personnel use. It includes all costs related to vehicles including vehicle insurance, fuel, parking fees, maintenance, repair, and registration taxes. Note that vehicle expenses incurred on construction projects need to be charged to those projects, not to general overhead costs.
Employee wages: This category includes the salaries and benefits of office and general management personnel. Executives, administration staff, and technical engineers working at your home

---

[1] In Chapter 1, we learned that overhead costs refer to costs that cannot be directly traced to cost objects in an economically feasible way.

office are included in this category. However, wages of employees working on a specific project would be charged to a project – that is a project overhead cost.

Miscellaneous general overhead costs: This includes legal fees, travel expenses, marketing and advertising costs, recruiting costs, charitable contributions, dues, and memberships.

As you can imagine, employee wages and office expenses (especially office rent or ownership costs) are major general overhead cost categories. Although each item of general overhead costs has its unique function or purpose, usually general overhead costs are said to contribute either to supporting management of existing contracts (i.e., existing projects) or to generating sales or new contracts.

4.2   Managing general overhead costs

4.2.1   Accurate general overhead allocation

Where do you use general overhead costs in your company? Why do you want to calculate or estimate general overhead costs? You may say, "We need to estimate general overhead costs to accurately calculate the estimated costs of the project on which we are placing a bid." Some may say, "We need to calculate general overhead costs to determine our profitability." The bottom line is that general overhead costs are needed to calculate business profit. Let us check how the profit of construction contractors can be calculated. As shown in Table 4.1, your gross profit is determined by subtracting your project costs (or construction costs) including project overhead costs from your revenue; your net profit from operation is determined by subtracting your general overhead costs from your gross profits. Note that you still need to consider other income, other expenses, and income tax to get your final net profit (not shown in Table 4.1).

Table 4.1   Income statement.

Revenue	(1)
Construction costs	(2)
Gross profit	(3) = (1) − (2)
General overhead costs	(4)
New profit from operation	(5) = (3) − (4)

As shown in Table 4.1, you can calculate your company's profit using your general overhead costs. However, many contractors allocate their general overhead costs to their projects; therefore, they have general overhead costs allocated to each project. With that information (i.e., general overhead costs allocated to each project), they can calculate net profit from operation on each project. Then, you may ask, why do you need to calculate the profit of each project? There are at least two reasons.

First, you need information on each project's profit to develop an effective marketing plan. Since most companies have limited resources, you need to focus your marketing resources on the most profitable market sectors and customers. How can you identify the profitable market sectors and customers without accurate cost data? Suppose that your company (construction contractor) has three market sectors: new commercial projects, renovation projects (commercial), and infrastructure (heavy civil) projects. Table 4.2 shows an example of profit analysis for each market sector when general overhead costs are allocated to projects based on revenue. According to Table 4.2, a new commercial construction is the least profitable sector while renovation and infrastructure sectors are the profitable sectors. Under this scenario, you may want to focus your marketing efforts on infrastructure and renovation projects.

Note, however, that general overhead costs are not necessarily consumed in proportion to revenue or contracted amount. Suppose that you have a small-sized renovation project where you need to deal with local regulations (of associated agencies) and complaints from neighbors. You can easily imagine that a small-sized project, in dealing with such issues, may consume a significant amount of your general overhead resources compared to the project's contracted amount. On the other hand, you may have a large infrastructure project in a remote area, one in which your home office staff do not have significant workload compared to their contracted amount. In that case, you may consume a minimal amount of general overhead resources on a large infrastructure project with a significant contracted amount.

Let's take a look at a different scenario, where you have accurate general overhead costs allocation (thanks to activity-based costing). Table 4.3 shows the result of profit analysis on each market sector with general overhead costs accurately allocated to projects. Net profit margin percentage on renovation projects in

Table 4.2 Income statement by market sectors (revenue-based allocation).

	Commercial, New	Commercial, Renovation	Infrastructure	Total
Revenue	$18,000,000.00	$7,300,000.00	$13,800,000.00	$39,100,000.00
Construction costs	$16,380,000.00	$6,533,500.00	$12,282,000.00	$35,195,500.00
Gross margin	$1,620,000.00	$766,500.00	$1,518,000.00	$3,904,500.00
Gross margin %	9.00%	10.50%	11.00%	9.99%
General overhead	$1,170,000.00	$474,500.00	$897,000.00	$2,541,500.00
Net profit from operation	$450,000.00	$292,000.00	$621,000.00	$1,363,000.00
Net profit margin %	2.50%	4.00%	4.50%	3.49%

Table 4.3 Income statement by market sectors (accurate overhead allocation).

	Commercial, New	Commercial, Renovation	Infrastructure	Total
Revenue	$18,000,000.00	$7,300,000.00	$13,800,000.00	$39,100,000.00
Construction costs	$16,380,000.00	$6,533,500.00	$12,282,000.00	$35,195,500.00
Gross margin	$1,620,000.00	$766,500.00	$1,518,000.00	$3,904,500.00
Gross margin %	9.00%	10.50%	11.00%	9.99%
General overhead	$1,118,160.00	$626,340.00	$797,000.00	$2,541,500.00
Profit from operation	$501,840.00	$140,160.00	$721,000.00	$1,363,000.00
Net profit margin %	2.79%	1.92%	5.22%	3.49%

Table 4.3 is lower as compared to the figure in Table 4.2 As you can see in Tables 4.2 and 4.3, a changed allocation of general overhead costs leads to different profit results in each market sector. The results may impact your marketing strategy. In other words, distorted general overhead allocation leads to the development of a misleading marketing strategy where marketing efforts may be focused on less profitable market sectors or customers.

Second, accurate general overhead allocation improves transparency in internal accountability (Grieco and Pilachiwshi, 1995). It has been found that many general contractors have multiple operational managers, each of whom is in charge of multiple construction projects. In many cases, operational managers are also in charge of the profitability of their projects. In other words, the profitability of projects allows you to evaluate their performances. However, you cannot evaluate the performance of operational managers fairly if you have distorted profit results on projects due to inaccurate allocation of general overheads.

4.2.2   *Providing a process view for process improvements*

Since the construction market is getting more and more competitive, reducing general overhead costs is key to making your business competitive. Reducing general overhead costs requires that you should use general overhead costs effectively. For that reason, many general contractors have undertaken process improvements. Movements or campaigns towards process efficiency or process improvement have been around in our industry for more than a decade. However, most process improvements that I have observed are one-time events, not sustainable efforts. Therefore, they are different from a system with a long-term approach or continuous improvement objectives. There is a gap between managerial accounting and process improvement efforts. In other words, information from managerial accounting does not provide any meaningful input to process improvement efforts. They are simply two separate and decoupled management tools.

To bridge the gap between managerial accounting and process improvement efforts, a function of managerial accounting needs to be to provide relevant cost information on major processes.

That being said, accounting information must provide management with a process view. You can calculate total general overhead costs simply by compiling all of the resource consumptions. However, the aggregate cost information (i.e., total general overhead costs) does not provide a process view, nor contribute to process improvements.

4.3 Does current practice for managing general overhead costs work?

The author carried out an industry survey to understand how practitioners in the construction industry manage their general overhead costs (Kim, 2002). General contractors' cost control reports were collected, and I carried out interviews with cost engineers and cost managers. Six (6) construction companies (general contractors) participated in the survey. The following observations were made based on the survey results.

4.3.1 Resource-based costing

Resource-based costing is defined in this book as a costing method in which overhead resources have individual cost accounts, and their costs are allocated to cost objects directly. Every company that participated in the survey used "resource-based costing," where resource costs are directly allocated to each project, in managing their general overhead costs.

It is hard to use cost data from resource-based costing in process improvements because it does not provide a process view. You may have experienced your company's process improvement efforts as just one-time events, instead of sustainable efforts. If you want process improvement efforts to be sustainable, you need to link managerial accounting functions with them. When your managerial accounting system provides cost information on processes, you can reduce efforts to acquire cost data for process improvement.

4.3.2 Volume-based assignment

The companies that participated in the survey assigned their general overhead costs to their individual projects. Four of the six participants assigned overhead costs in proportion to direct labor hours or direct labor costs while two of them assigned overhead costs in proportion to the contract amounts of each project. The following example shows how general overhead costs can be assigned by volume-based assignment.

EXAMPLE

Suppose that XYZ Construction, Inc., a general contractor, has 'n' projects (e.g., $i = 1, 2, 3, ..., n$). Suppose that contracted amounts for those projects are A_1, A_2, A_3, ..., and A_n, respectively. Direct labor costs for those projects are L_1, L_2, L_3, ..., and L_n, respectively. Given that total general overhead costs of XYZ Construction for a certain period are X, the general overhead costs assigned to each project $(X_1, X_2, X_3, ...,$ and $X_n)$ can be calculated in two ways: contract amount-based allocation and direct labor costs-based allocation.

If general overhead costs are allocated based on contracted amount or revenue for each project, then general overhead costs will be allocated as follows:

$$Xi = X \times Ai / \sum\nolimits_{i=1}^{n} Ai$$

where $i = 1, 2, ..., n$

If general overhead costs are allocated based on direct labor costs, then overhead costs will be allocated as follows:

$$Xi = X \times Li / \sum\nolimits_{i=1}^{n} Xi$$

where $i = 1, 2, ..., n$

All of us recognize that general overhead costs are not necessarily in proportion to direct labor costs or contract amounts, as shown in Tables 4.2 and 4.3. One of the problems with current practices is that companies do not know the real costs of their projects either because they do not assign general overhead costs to their projects or because they use a volume-based allocation such as direct labor costs or revenue of each project in assigning general overhead costs to projects. As a result, it is difficult to find out where money is being earned and lost. Inaccurate allocation of general overhead cost to projects leads to cost distortion. In some cases, a project that generates profits can be red-flagged as a "loser," which may lead to a poor strategic plan.

4.4 How can ABC be implemented in managing general overhead costs?

In previous chapters, we discussed the general methodology of the ABC system (Chapter 2), and a case where ABC was applied to managing project overhead costs (Chapter 3). In this chapter, we will examine a case where ABC is applied to managing general overhead costs.

4.4.1 Case study: xx Construction (general contractor)

This case study details a real organization where the author worked as a consultant, although the numbers are modified. In this case, the general contractor has three market sectors: (1) new commercial, (2) renovation, and (3) infrastructure. In the past, the contractor focused on the new commercial construction sector. Recently the company's infrastructure sector has expanded due to increased government spending on infrastructure. The income statement of the contractor is to be found in Table 4.4.

Management wanted to apply ABC to managing their general overhead costs for the following reasons:

1) They wanted to understand the true profit structure through accurate allocation of their general overhead costs to projects. They needed profit information on each project because they wanted to focus their marketing strategy on the most profitable market sector(s).

Table 4.4   Income statement.

Period: June 201x		
Revenue		**$16,100,000.00**
	Tenant improvement	$5,100,000.00
	New commercial	$8,000,000.00
	Infrastructure	$3,000,000.00
Construction costs		**$15,008,000.00**
	Tenant improvement	$4,743,000.00
	New commercial	$7,520,000.00
	Infrastructure	$2,745,000.00
Gross margin		**$1,092,000.00**
General overhead		$676,200.00
Net profit from operation		**$415,800.00**

2) They wanted to reduce general overhead costs, especially human resource costs at their home office, because the construction market is getting more competitive. Even a small reduction in general overhead costs can make a difference in financial performance.

As a first step toward reducing general overhead costs, management wanted to acquire cost information on major processes to be used for process improvements.

4.4.1.1 Preliminary investigation of general overhead costs

The accounting department made a preliminary investigation of their general overhead costs to identify major components of general overhead costs before taking on ABC. As shown in Table 4.5, the company's general overhead costs consisted of four major functions: (1) preconstruction, (2) project management and support, (3) organization-sustaining, and (4) facilities, utilities, and supplies.

4.4.1.2 Developing an activity-based costing charter

As seen in the example of a project-based ABC system (Figure 3.2 in the previous chapter), an ABC charter needs to be developed after a task force is established. The ABC charter contributes to developing a consensus within your organization. In this case, the task force was made up of five individuals including an external facilitator. The team developed a charter, which included the following components:

- *Scope*: The team preferred to limit the scope of general overhead costs in the ABC system to the project management function. Therefore, general overhead costs in the area of

Table 4.5   Major components of general overhead costs.

General Overhead		$676,200.00
Preconstruction	16%	$108,192.00
Project management/support	47%	$317,814.00
Organization-sustaining	15%	$101,430.00
Facility, utility, supplies	22%	$148,764.00

Table 4.6   Scope of ABC.

General overhead cost category	ABC system	Allocation base
Project management & support	Included	Cost drivers
Preconstruction	Not included	Revenue or contracted amount
Organization-sustaining	Not included	Revenue or contracted amount
Facility, utility, supplies	Not included	Revenue or contracted amount

project management need to be investigated and analyzed into activity costs, which are then to be allocated to cost objects using cost drivers. On the other hand, other general overhead costs (i.e., preconstruction, organization-sustaining, and facility/utilities costs) are to be allocated to cost objects based on the revenue of each cost object (Table 4.6).

- *Objectives*: As mentioned, the ABC system in this case had two goals: (1) to obtain data on accurate profit for each project and each market sector through accurate allocation of general overhead costs, and (2) identification of major processes with their costs. The team developed four specific objectives to achieve these two goals:
  1) To identify major activities
  2) To estimate activity costs
  3) To identify cost driver rates
  4) To calculate the profit ratio (%) of each project and each market sector.
- *Logistics*: The task force decided to have weekly meetings while developing their ABC system. It is important to get assistance and support from employees, especially when collecting data such as activity information. The company set up a two-hour workshop in which employees working in the associated departments (sustainability, quality/safety/health, accounting/finance, and procurement) were instructed on how data would be collected and used.

A sample team charter for the ABC system to manage general overhead costs is to be found in Figure 4.1.

ACTIVITY-BASED COSTING CHARTER Home office overhead	
SCOPE	
1. Human resources on project management and support	
2. NOT INCLUDED: preconstruction, organization-sustaining, facility/utility/supplies	
TEAM MEMBERS	
Executive sponsor:	xxxx (Vice President)
1. Facilitator:	xxxx
2. Team members:	xxxx
OBJECTIVES	
1. Identify major activities	
2. Identify activity costs	
3. Identify cost driver rates	
4. Calculate profit % of each project and each market sector	
LOGISTICS	
Regular Meeting	7:00 AM–9:00 AM on Friday
Workshop	2 hours; all project members are expected to participate
Report	by 5:00 PM December x, 20xx
AGREEMENT	
Team members' signature	
Vice president's signature	

Figure 4.1   **Sample ABC team charter for managing general overhead costs.**

4.4.1.3   Defining cost objects

The task force defined three types of cost objects: (1) projects, (2) market sectors, and (3) customers. In effect, the team had only one type of cost object: projects. This choice was made because cost and profit information on both market sectors and customers is just reconfiguration of cost and profit information on projects.

Table 4.7   Major functions and activity centers.

Activity center category	Activity center
Preconstruction	
	Design consulting
	Estimating
	Bidding
	Scheduling
Project management	
	Sustainability (green)
	Quality/health/safety
	Accounting/finance
	Procurement
Organization-sustaining	
	Human resources
	Marketing
	General admin
Facilities, utilities, supplies	Not included in activity analysis

4.4.1.4   Identifying activities

As mentioned, four functions were selected: (1) preconstruction, (2) project management/support, (3) organization-sustaining, and (4) facility/utility/supplies. Each function had multiple activity centers, as shown in Table 4.7.

The task force interviewed key employees of each department in project management/support functions to identify associated activities. Note that the team decided to include only project management/support functions in the ABC system (Table 4.7). The list of activities is provided in Table 4.8.

It is convenient to use activity codes in managing your ABC system if you have a large number of activities and activity centers as seen in this case. Note that the project-based ABC case study (Chapter 3) did not use an activity code system because the number of activity centers and activities were limited. Activity codes consisting of the abbreviation of each function, activity center number, and activity number were used to identify each activity account in this case. The activity code was formatted as follows:

$$\#\# \text{-} \#\# \text{-} \#\#$$

where the two letters to the left of the first hyphen represent the function of general overhead resources (e.g., project management/support, preconstruction), the two digit number between

Table 4.8   List of activities.

Activity Center: Sustainable Construction	
PM-01-01	Green construction training
PM-01-02	Design charrette consulting
PM-01-03	LEED documentation (draft)
PM-01-04	LEED updating
PM-01-05	LEED communication
Activity Center: Health/Safety/Quality	
PM-02-01	Training
PM-02-02	Site manual review/approval
PM-02-03	Site manual updates
PM-02-04	Site auditing
PM-02-05	Auditing follow-up
Activity Center: Accounting/Finance	
PM-03-01	Budget approval
PM-03-02	Budget updates (CO)
PM-03-03	CO request reviewed
PM-03-04	Progress payment req. review
PM-03-05	Monthly project accounting review/approval
PM-03-06	Monthly accounting closing
PM-03-07	Book keeping
Activity Center: Procurement	
PM-04-01	Subcontract review/approval
PM-04-02	Subcontract draft
PM-04-03	Contract draft
PM-04-04	Contract revision

the hyphens represents the activity center, and the last two digit number represents the activity. For example, an activity of "subcontract draft" is coded as PM-04–02 in Table 4.8. The first two letters, PM, represent the function "project management/support." The two digit number between hyphens, 04, represents the activity center "procurement." The last two digit number, 02, represents an activity of "subcontract draft."

4.4.1.5   Assigning resource costs to activities

The team chose to use time–effort % method, where each person's time is allocated to each activity because it is simple and easy to update compared to tracking actual time spent on each activity for each person.

The team held a survey in which each employee was required to allocate the percentage (%) of his or her time spent on each

Table 4.9   Example of survey results (time–effort % method).

	Training	Design charrette consulting	LEED draft	LEED updating	LEED communication	Total percentage
Adams	20%	30%	25%	15%	10%	100%
Kollman	0%	0%	30%	30%	40%	100%
Cox	0%	0%	30%	40%	30%	100%
Anderson	0%	20%	30%	25%	25%	100%
Kim	30%	40%	10%		20%	100%
Bender	20%	20%	20%	10%	30%	100%
Bailey	70%	30%	0%			100%

activity in four departments (sustainable construction, quality/ health/safety, accounting/finance, and procurement) corresponding to the four activity centers. Table 4.9 shows an example of the survey results for the activity center sustainability.

Since each department was in charge of each activity center, estimating the costs of each activity center was relatively straightforward. Cost information on each department was provided by the accounting department. Therefore, each department's costs were allocated to activities identified in Table 4.8 according to the survey results. The result of activity costs allocation is shown in Table 4.10.

4.4.1.6   Determining a cost driver for each activity

As in the activity-based costing project case study (Chapter 3), the following criteria were used to test if a particular factor can be used as a cost driver. The team used three criteria in choosing a cost driver for each activity:

- Does it have a cause–effect relationship with a cost object? Does the cost of a cost object increase in proportion to the volume of a cost driver?
- Can we measure the volume of a cost driver in an objective way? If you use the level of customer satisfaction as a cost driver, for example, you may have hard time in measuring the volume of the cost driver (i.e., customer satisfaction) in an objective way.
- Can we measure the volume of a cost driver in an economically feasible way? Your ABC system requires regular updating, which includes measuring the volume of a cost driver on a regular basis, since this is not constant over time.

Table 4.10   List of cost drivers.

Activity ID	Activity	Costs
Sustainable Construction		
PM-01-01	Green construction training	$12,895.00
PM-01-02	Design charrette consulting	$14,070.00
PM-01-03	LEED documentation (draft)	$16,165.00
PM-01-04	LEED updating	$13,417.50
PM-01-05	LEED communication	$16,552.50
Health/Safety/Quality		
PM-02-01	Training	$29,400.00
PM-02-02	Site manual review/approval	$18,450.00
PM-02-03	Site manual updates	$13,000.00
PM-02-04	Site auditing	$25,400.00
PM-02-05	Auditing follow-up	$12,000.00
Accounting/Finance		
PM-03-01	Budget approval	$3,300.00
PM-03-02	Budget updates (CO)	$9,760.00
PM-03-03	CO request reviewed	$11,316.00
PM-03-04	Progress payment req. review	$9,920.00
PM-03-05	Monthly project accounting review/approval	$15,100.00
PM-03-06	Monthly closing	$19,800.00
PM-03-07	Book keeping	$16,528.00
Procurement		
PM-04-01	Subcontract review/approval	$19,860.00
PM-04-02	Subcontract draft	$17,450.00
PM-04-03	Contract draft	$13,670.00
PM-04-04	Contract revision	$9,760.00

Using the above criteria, the task force held two meetings to set up cost drivers. The following section addresses how the team chose cost drivers. The complete set of cost drivers is to be found in Table 4.11.

Activity Center #1: Sustainable construction

The team decided to use a duration cost driver for the training and design charrette consulting activities. The number of training hours was used on the "green construction training" activity for sustainable construction. There was discussion about whether the number of trainees could be used as a cost driver. But the team

Table 4.11   List of activity costs.

Cost Driver & Activity
Duration cost driver	*Sustainable Construction* # of training hours (Training) # of charrette hours (Design consulting) *Health/Safety/Quality* # of training hours (Training) # of review hours (Site manual review) # of auditing hours (Site auditing) *Accounting/Finance* # of review hours (Progress payment review) *Procurement* # of draft hours (Subcontract draft) # of draft hours (Contract draft)
Transactional cost driver	*Sustainable Construction* # of credits (LEED draft; LEED updates) # of questions (LEED communication) # of items (Site manual updates) *Safety/Health/Quality* # of failures (Auditing follow-up) *Accounting/Finance* # of pages (Budget approval) # of change order items approved (Budget updates) # of change orders reviewed (Change order request review) *Procurement* # of revisions (Subcontract revision) # of revisions (Contract revision)
Budget cost driver	# of construction cost ratio (Monthly project report)

rejected this idea because the "green construction training" activity does not cost double when the number of trainees in a classroom is doubled. Similarly, the number of charrette hours was used as a cost driver for the activity of "design charrette consulting."

On the other hand, a transactional cost driver was used on activities associated with LEED[2] documentation such as drafting, updating, and communication with USGBC (US Green Building

---

[2] LEED (Leadership in Energy and Environmental Design).

Council), an accreditation agency of LEED. The activities of "LEED drafting" and "LEED revising" used the number of credits as their cost driver. Although there was discussion about whether the number of points could be used as a cost driver, the team rejected the idea because the activities associated with LEED documents do not cost twice as much when the assigned points on a credit are doubled. For example, the "Credit 4.2 Alternative Transportation – Bicycle Storage and Changing Rooms" in a section on "sustainable sites" has six points while the "Credit 4.1 Alternative Transportation – Public Transportation Access" in the same section has only one point (USGBC 2013). You do not expect that activities on Credit 4.2 cost six times as much as activities on Credit 4.1 do.

Activity Center #2: Health, Safety, and Quality

Similar to the activity of "sustainability training" (PM-01–01), a duration cost driver (i.e., the number of training hours) was used as a cost driver for the activity of health/safety/quality training (PM-02–01). A duration cost driver was also used as a cost driver for the activity of "site manual review and approval" (PM-02–02) although some members of the task force argued that a transactional cost driver might be used. Through discussions on the estimated time spent on reviewing site manuals, the team rejected the idea of using a transactional cost driver, because the time spent reviewing items in a site manual would fluctuate significantly. Similarly, the activity of "site auditing" (PM-02–04) used a duration driver, the number of auditing hours.

On the other hand, the activity of "site manual updates" used a transactional cost driver, the number of items updated. Similarly, the activity of "auditing follow-up" (PM-02–05) used a transactional cost driver, namely the number of failures.

Activity Center #3: Accounting/finance

A transactional cost driver (i.e., the number of pages, the number of change order items) was used for the activities associated with project budget and change order requests. On the other hand, the activities associated with accounting and progress payment review used a duration cost driver (i.e., the number of hours).

Activity Center #4: Procurement

The activities of "subcontract draft/review" (PM-04–02) and "contract draft" (PM-04–03) used a duration cost driver (i.e., the number of hours) because the duration of such activities fluctuates. On the other hand, the activities of "subcontract revision" (PM-04–01) and "contract revision" used a transactional cost driver (i.e., the number of revisions).

4.4.1.7   Calculating the unit rate of activity costs (activity-cost rate) and allocating activity costs to cost objects

You need to calculate the unit rate of activity costs (i.e., activity cost/total volume of a cost driver) by measuring the total volume of each cost driver. When you measure the volume of cost drivers, you need to record them on each cost object. Although the total volume of a cost driver is needed in calculating the activity-cost rates, the volume of a cost driver on each cost object is eventually required when allocating the activity costs to cost objects.

You can get cost driver information (i.e., the volume of each cost driver) through examining relevant documents in the case of a transactional cost driver. For example, the number of change orders can be counted by examining documents relevant to change orders in the case of the "budget update" activity. You do not have to ask someone in charge or use a survey. However, you need to ask someone in charge or use a survey to estimate the volume of a duration cost driver such as time spent on a specific activity.

Table 4.12 shows the unit rates of activity costs and the volume of cost drivers on each cost object (i.e., each project); Tables 4.13 and 4.14 present the volume of cost drivers on each cost object and the costs of cost objects (i.e., the result of cost allocation).

4.5   How can ABC data be used in managing general overhead costs?

There were four objectives set up for this ABC project: (1) to identify major activities, (2) to estimate activity costs, (3) to calculate the activity-cost rates, and (4) to calculate the profit percentage of each project and each market sector. The first and second objectives were achieved in the course of ABC system development. In this section, we will discuss how ABC data can be used to meet other objectives.

Table 4.12   Unit rates of activity costs.

Activity	Activity costs	Cost driver volume	Activity-cost rates
Sustainable Construction			
Training	$12,895.00	85	$151.71
Design charrette consulting	$14,070.00	40	$351.75
LEED documentation (draft)	$16,165.00	190	$85.08
LEED updating	$13,417.50	85	$157.85
LEED communication	$16,552.50	37	$447.36
Health/Safety/Quality			
Training	$29,400.00	75	$392.00
Site manual review/approval	$19,450.00	55	$353.64
Site manual updates	$12,000.00	20	$600.00
Site auditing	$25,400.00	28	$907.14
Auditing Follow-up	$12,000.00	24	$500.00
Accounting/Finance			
Budget approval	$3,300.00	93	$35.48
Budget updates (CO)	$9,760.00	15	$650.67
CO request reviewed	$11,316.00	30	$377.20
Progress payment req. review	$9,920.00	27	$367.41
Monthly project accounting review/appr	$15,100.00	Cost ratio	
Monthly closing	$19,800.00	Cost ratio	
Book keeping	$16,528.00	Cost ratio	
Procurement			
Subcontract review/approval	$19,860.00	9	$2,206.67
Subcontract draft	$17,450.00	40	$436.25
Contract draft	$13,670.00	28	$488.21
Contract revision	$9,760.00	3	$3,253.33

4.5.1   Cost driver analysis

The unit rate of activity costs (or cost driver rate) can be used to identify activities for process improvements. The cost driver rates are to be found in Table 4.15.

4.5.2   Profitability analysis for each project

The accurate allocation of general overhead costs to cost objects allows you to accurately calculate the profitability of each

Table 4.13 Volume of cost drivers on each cost object.

	Project 1A	Project 1B	Project 1C	Project 2A	Project 2B	Project 2C	Project 2D	Project 3A	Project 3B	Total
Sustainable Construction										
Training		15	13	10	12	15		10	10	85
Design charrette consulting		11	8	8	6	7				40
LEED documentation (draft)		40	36	38	40	36				190
LEED updating	8	12	15	14	16	13	7			85
LEED communication	5	9	6	5	5	4	3			37
Health/Safety/Quality										
Training	9	10	12	6	10	10	8	10		75
Site manual review/approval		8	10	6	8	8		8	7	55
Site manual updates	2	3	3	1	3	4	1	3		20
Site auditing	7		5	5		5		6		28
Auditing Follow-up	6		6	4		5		3		24
Accounting/Finance										
Budget approval		15	15	13	14	14		11	11	93
Budget updates (CO)	3	2	3	2	1	2		2		15

CO request reviewed	5	4	6	4	2	3	2	3	1	30
Progress payment req. review	3	3.5	3	3	3	3	3	3	2.5	27
Monthly project accounting review	0.12	0.11	0.09	0.11	0.14	0.12	0.13	0.11	0.07	1
Monthly closing	0.12	0.11	0.09	0.11	0.14	0.12	0.13	0.11	0.07	1
Book keeping	0.12	0.11	0.09	0.11	0.14	0.12	0.13	0.11	0.07	1
Procurement										
Subcontract review/revision	2	1	2	1		2	1			9
Subcontract draft	6	4	8	4	6	5		4	3	40
Contract draft		4	4	4	4	4		4	4	28
Contract revision		1	1	1						3

Table 4.14 Costs of cost objects.

	Project 1A	Project 1B	Project 1C	Project 2A	Project 2B	Project 2C	Project 2D	Project 3A	Project 3B	Total
Sustainable Construction										
Training	$-	$2,275.59	$1,972.18	$1,517.06	$1,820.47	$2,275.59	$-	$1,517.06	$1,517.06	$12,895.00
Design charrette consulting	$-	$3,869.25	$2,814.00	$2,814.00	$2,110.50	$2,462.25	$-	$-	$-	$14,070.00
LEED documentation (draft)	$-	$3,403.16	$3,062.84	$3,233.00	$3,403.16	$3,062.84	$-	$-	$-	$16,165.00
LEED updating	$1,262.82	$1,894.24	$2,367.79	$2,209.94	$2,525.65	$2,052.09	$1,104.97	$-	$-	$13,417.50
LEED communication	$2,236.82	$4,026.28	$2,684.19	$2,236.82	$2,236.82	$1,789.46	$1,342.09	$-	$-	$16,552.50
Health/Safety/Quality										$-
Training	$3,528.00	$3,920.00	$4,704.00	$2,352.00	$3,920.00	$3,920.00	$3,136.00	$3,920.00	$-	$29,400.00
Site manual review/approval	$-	$2,829.09	$3,536.36	$2,121.82	$2,829.09	$2,829.09	$-	$2,829.09	$2,475.45	$19,450.00
Site manual updates	$1,200.00	$1,800.00	$1,800.00	$600.00	$1,800.00	$2,400.00	$600.00	$1,800.00	$-	$12,000.00
Site auditing	$6,350.00	$-	$4,535.71	$4,535.71	$-	$4,535.71	$-	$5,442.86	$-	$25,400.00
Auditing follow-up	$3,000.00	$-	$3,000.00	$2,000.00	$-	$2,500.00	$-	$1,500.00	$-	$12,000.00
Accounting/Finance										$-
Budget approval	$-	$532.26	$532.26	$461.29	$496.77	$496.77	$-	$390.32	$390.32	$3,300.00
Budget updates (CO)	$1,952.00	$1,301.33	$1,952.00	$1,301.33	$650.67	$1,301.33	$-	$1,301.33	$-	$9,760.00
CO request reviewed	$1,886.00	$1,508.80	$2,263.20	$1,508.80	$754.40	$1,131.60	$754.40	$1,131.60	$377.20	$11,316.00
Progress payment req. review	$1,102.22	$1,285.93	$1,102.22	$1,102.22	$1,102.22	$1,102.22	$1,102.22	$1,102.22	$918.52	$9,920.00
Monthly project accounting review	$1,750.92	$1,615.80	$1,405.35	$1,688.54	$2,055.40	$1,864.16	$1,958.00	$1,638.04	$1,123.79	$15,100.00
Monthly closing	$2,295.91	$2,118.73	$1,842.78	$2,214.11	$2,695.16	$2,444.39	$2,567.44	$2,147.89	$1,473.58	$19,800.00
Book keeping	$1,916.51	$1,768.61	$1,538.25	$1,848.22	$2,249.78	$2,040.45	$2,143.17	$1,792.94	$1,230.07	$16,528.00
Procurement										$-
Subcontract review/revision	$4,413.33	$2,206.67	$4,413.33	$2,206.67	$-	$4,413.33	$2,206.67	$-	$-	$19,860.00
Subcontract draft	$2,617.50	$1,745.00	$3,490.00	$1,745.00	$2,617.50	$2,181.25	$-	$1,745.00	$1,308.75	$17,450.00
Contract draft	$-	$1,952.86	$1,952.86	$1,952.86	$1,952.86	$1,952.86	$-	$1,952.86	$1,952.86	$13,670.00
Contract revision	$-	$3,253.33	$3,253.33	$3,253.33	$-	$-	$-	$-	$-	$9,760.00
TOTAL	$35,512.05	$43,306.92	$54,222.67	$42,902.73	$35,220.46	$46,755.40	$16,914.96	$30,211.21	$12,767.60	$317,814.00

Table 4.15   Cost driver analysis.

Cost per training hour	
Sustainable construction training	$198.38
Health/safety/quality training	$392.00
Cost per LEED credit	
LEED documentation (draft)	$67.35
LEED updating	$111.81
Cost per change order item	
Budget updates (CO)	$305.00
Change order request reviewed	$188.60
Cost per contract revision	
Subcontract revision	$902.73
Contract revision	$813.33

project as opposed to the case where general overhead costs are allocated based on revenue or construction costs. Table 4.16 shows an income schedule on projects using traditional volume-based allocation while Table 4.17 shows an income schedule using ABC.

Table 4.18 shows how much the net profit ratio on each project changes when the overhead allocation method is changed to ABC. Although ABC analysis was applied only to the project management sector, the net profit ratios on projects change as much as 1.74% (Project 1C) as shown in Table 4.18. You may think that 1.74% is a small change. However, that change is not small when the fact that the industry average net profit ratio is usually less than five to six percent is taken into account.

4.5.3 *Profitability analysis for each market sector*

Profitability analysis results are usually used not only for marketing purposes but also for accountability. First, companies need to focus their marketing efforts on profitable sectors, due to limited resources in many cases. As the number of projects using negotiated contracts or qualification-based selection (i.e., non-price competition) grows, marketing has become a competitive edge in the construction industry. Experienced construction contractors have marketing people whose primary focus is building and nurturing relationships with potential customers (Jackson, 2011). If your company is able to identify profitable customers, your marketing efforts can be more effective than those of your competitors.

Table 4.16 Income schedule using a traditional volume-based allocation.

	Project 1A	Project 1B	Project 1C	Project 2A	Project 2B	Project 2C	Project 2D	Project 3A	Project 3B
Revenue	$1,904,000.00	$1,776,500.00	$1,419,500.00	$1,750,000.00	$2,240,000.00	$1,920,000.00	$2,090,000.00	$1,815,000.00	$1,185,000.00
Construction costs	$1,740,256.00	$1,605,956.00	$1,396,788.00	$1,678,250.00	$2,042,880.00	$1,852,800.00	$1,946,070.00	$1,628,055.00	$1,116,945.00
Gross margin	$163,744.00	$170,544.00	$22,712.00	$71,750.00	$197,120.00	$67,200.00	$143,930.00	$186,945.00	$68,055.00
Gross margin %	8.60%	9.60%	1.60%	4.10%	8.80%	3.50%	6.89%	10.30%	5.74%
General overhead	$78,408.92	$72,357.91	$62,933.64	$75,615.18	$92,043.94	$83,479.70	$87,682.07	$73,353.60	$50,325.04
Preconstruction	$12,545.43	$11,577.26	$10,069.38	$12,098.43	$14,727.03	$13,356.75	$14,029.13	$11,736.58	$8,052.01
Project management	$36,852.19	$34,008.22	$29,578.81	$35,539.14	$43,260.65	$39,235.46	$41,210.57	$34,476.19	$23,652.77
Organization-sustaining	$11,761.34	$10,853.69	$9,440.05	$11,342.28	$13,806.59	$12,521.96	$13,152.31	$11,003.04	$7,548.76
Facility, utility, supplies	$17,249.96	$15,918.74	$13,845.40	$16,635.34	$20,249.67	$18,365.53	$19,290.06	$16,137.79	$11,071.51
Net profit from operation	$85,335.08	$98,186.09	$(40,221.64)	$(3,865.18)	$105,076.06	$(16,279.70)	$56,247.93	$113,591.40	$17,729.96
Net profit %	4.48%	5.53%	-2.83%	-0.22%	4.69%	-0.85%	2.69%	6.26%	1.50%

Table 4.17 Income schedule using ABC.

	Project 1A	Project 1B	Project 1C	Project 2A	Project 2B	Project 2C	Project 2D	Project 3A	Project 3B
Revenue	$1,904,000.00	$1,776,500.00	$1,419,500.00	$1,750,000.00	$2,240,000.00	$1,920,000.00	$2,090,000.00	$1,815,000.00	$1,185,000.00
Construction costs	$1,740,256.00	$1,605,956.00	$1,396,788.00	$1,678,250.00	$2,042,880.00	$1,852,800.00	$1,946,070.00	$1,628,055.00	$1,116,945.00
Gross margin	$163,744.00	$170,544.00	$22,712.00	$71,750.00	$197,120.00	$67,200.00	$143,930.00	$186,945.00	$68,055.00
Gross margin %	8.60%	9.60%	1.60%	4.10%	8.80%	3.50%	6.89%	10.30%	5.74%
General overhead	$77,068.78	$81,656.61	$87,577.49	$82,978.78	$84,003.75	$90,999.65	$63,386.46	$69,088.61	$39,439.87
Preconstruction	$12,545.43	$11,577.26	$10,069.38	$12,098.43	$14,727.03	$13,356.75	$14,029.13	$11,736.58	$8,052.01
Project management	$35,512.05	$43,306.92	$54,222.67	$42,902.73	$35,220.46	$46,755.40	$16,914.96	$30,211.21	$12,767.60
Organization-sustaining	$11,761.34	$10,853.69	$9,440.05	$11,342.28	$13,806.59	$12,521.96	$13,152.31	$11,003.04	$7,548.76
Facility, utility, supplies	$17,249.96	$15,918.74	$13,845.40	$16,635.34	$20,249.67	$18,365.53	$19,290.06	$16,137.79	$11,071.51
Net profit from operation	$86,675.22	$88,887.39	$(64,865.49)	$(11,228.78)	$113,116.25	$(23,799.65)	$80,543.54	$117,856.39	$28,615.13
Net profit %	4.55%	5.00%	−4.57%	−0.64%	5.05%	−1.24%	3.85%	6.49%	2.41%

Table 4.18 Changes in net profit ratio, projects.

Project	Type	Operational margin %, traditional	Operational margin %, ABC	Change %
Project 1A	Commercial, renovation	4.48%	4.55%	0.07%
Project 1B	Commercial, renovation	5.53%	5.00%	−0.52%
Project 1C	Commercial, renovation	−2.83%	−4.57%	−1.74%
Project 2A	Commercial, new	−0.22%	−0.64%	−0.42%
Project 2B	Commercial, new	4.69%	5.05%	0.36%
Project 2C	Commercial, new	−0.85%	−1.24%	−0.39%
Project 2D	Commercial, new	2.69%	3.85%	1.16%
Project 3A	Infrastructure	6.26%	6.49%	0.23%
Project 3B	Infrastructure	1.50%	2.41%	0.92%

Table 4.19 Income schedule using volume-based allocation, market sector.

Cost object: market sector (traditional method, construction cost-based allocation)	Tenant improvements	New commercial	Infrastructure
Revenue	$5,100,000.00	$8,000,000.00	$3,000,000.00
Construction costs	$4,743,000.00	$7,520,000.00	$2,745,000.00
Gross margin	$357,000.00	$480,000.00	$255,000.00
Gross margin %	7.00%	6.00%	8.50%
General overhead	$213,700.47	$338,820.90	$123,678.64
Preconstruction	$34,192.07	$54,211.34	$19,788.58
Project management	$100,439.22	$159,245.82	$58,128.96
Organization-sustaining	$32,055.07	$50,823.13	$18,551.80
Facility, utility, supplies	$47,014.10	$74,540.60	$27,209.30
Net profit from operation	$143,299.53	$141,179.10	$131,321.36
Net profit %	2.81%	1.76%	4.38%

Table 4.20   Income schedule using ABC, market sector.

Cost object : individual projects (activity-based costing)			
	Tenant improvements	New commercial	Infrastructure
Revenue	$5,100,000.00	$8,000,000.00	$3,000,000.00
Construction costs	$4,743,000.00	$7,520,000.00	$2,745,000.00
Gross margin	$357,000.00	$480,000.00	$255,000.00
Gross margin %	7.00%	6.00%	8.50%
General overhead	$246,302.88	$321,368.63	$108,528.49
Preconstruction	$34,192.07	$54,211.34	$19,788.58
Project management	$133,041.63	$141,793.56	$42,978.81
Organization-sustaining	$32,055.07	$50,823.13	$18,551.80
Facility, utility, supplies	$47,014.10	$74,540.60	$27,209.30
Net profit from operation	$110,697.12	$158,631.37	$146,471.51
Net profit %	2.17%	1.98%	4.88%

Table 4.21   Profit % changes, market sector.

Market sector	Operational margin %, traditional	Operational margin %, ABC	Net profit change
Commercial, renovation	2.81%	2.17%	−0.64%
Commercial, new	1.76%	1.98%	0.22%
Infrastructure	4.38%	4.88%	0.51%

Second, companies usually have management personnel in charge of each market sector. Accurate profit analysis of each market sector resulting from accurate allocation of general overhead costs provides a fair assessment tool to evaluate them.

Once you have income statements on projects with general overhead properly allocated, you can easily generate income statements on market sectors. For example, profits on commercial renovation are the summation of profits on projects 1A, 1B, and 1C. Table 4.19 presents the income statement on each market sector using a traditional volume-based allocation while Table 4.20 shows the income statement using ABC.

Table 4.22   Income schedule using volume-based allocation, customer.

Cost object: customer (traditional method, construction cost-based allocation)						
	University of xx	Micxxx	Vulxxx	Avaxxx	xDOT	City of x
Revenue	$1,750,000.00	$3,680,500.00	$3,659,500.00	$4,010,000.00	$1,815,000.00	$1,185,000.00
Construction costs	$1,678,250.00	$3,346,212.00	$3,439,668.00	$3,798,870.00	$1,628,055.00	$1,116,945.00
Gross margin	$71,750.00	$334,288.00	$219,832.00	$211,130.00	$186,945.00	$68,055.00
Gross margin %	4.10%	9.08%	6.01%	5.27%	10.30%	5.74%
General overhead	$75,615.18	$150,766.83	$154,977.58	$171,161.77	$73,353.60	$50,325.04
Preconstruction	$12,098.43	$24,122.69	$24,796.41	$27,385.88	$11,736.58	$8,052.01
Project management	$35,539.14	$70,860.41	$72,839.46	$80,446.03	$34,476.19	$23,652.77
Organization-sustaining	$11,342.28	$22,615.02	$23,246.64	$25,674.27	$11,003.04	$7,548.76
Facility, utility, supplies	$16,635.34	$33,168.70	$34,095.07	$37,655.59	$16,137.79	$11,071.51
Net profit from operation	$-3,865.18	$183,521.17	$64,854.42	$39,968.23	$113,591.40	$17,729.96
Net profit %	−0.22%	4.99%	1.77%	1.00%	6.26%	1.50%

Table 4.23 Income schedule using ABC, market sector, customer.

Cost object: customer (activity-based costing)						
	University of xx	Micxxx	Vulxxx	Avaxxx	xDOT	City of x
Revenue	$1,750,000.00	$3,680,500.00	$3,659,500.00	$4,010,000.00	$1,815,000.00	$1,185,000.00
Construction costs	$1,678,250.00	$3,346,212.00	$3,439,668.00	$3,798,870.00	$1,628,055.00	$1,116,945.00
Gross margin	$71,750.00	$334,288.00	$219,832.00	$211,130.00	$186,945.00	$68,055.00
Gross margin %	4.10%	9.08%	6.01%	5.27%	10.30%	5.74%
General overhead	$82,978.78	$158,725.39	$171,581.24	$154,386.11	$69 088.61	$39,439.87
Preconstruction	$12,098.43	$24,122.69	$24,796.41	$27,385.88	$11.736.58	$8,052.01
Project management	$42,902.73	$78,818.97	$89,443.13	$63,670.37	$30,211.21	$12,767.60
Organization-sustaining	$11,342.28	$22,615.02	$23,246.64	$25,674.27	$11,003.04	$7,548.76
Facility, utility, supplies	$16,635.34	$33,168.70	$34,095.07	$37,655.59	$1€,137.79	$11,071.51
Net profit from operation	$-11,228.78	$175,562.61	$48,250.76	$56,743.89	$117,856.39	$28,615.13
Net profit %	−0.64%	4.77%	1.32%	1.42%	6.49%	2.41%

Table 4.24   Profit % changes, customer.

Customers	Operational margin %, traditional	Operational margin %, ABC	Net profit change
University of xx	−0.22%	−0.64%	−0.42%
Micxxx	4.99%	4.77%	−0.22%
Vulxxx	1.77%	1.32%	−0.45%
Avaxxx	1.00%	1.42%	0.42%
xDOT	6.26%	6.49%	0.23%
City of xx	1.50%	2.41%	0.91%

Table 4.21 shows how much the net profit rate on each market sector changes when the overhead allocation method changes to ABC. According to Table 4.21, the net profit percentage on renovation projects decreased while the net profit percentage on new commercial and infrastructure increased. Note that the result in Table 4.21 shows only one-month's data. You may want to use more comprehensive results such as profitability analysis results over more than a year for marketing or strategic planning purposes.

4.5.4   Profitability analysis for each customer

With the results of profitability analysis for each customer, contractors can focus their marketing efforts on profitable customers. The calculation of profitability analysis for each customer is straightforward once you have profitability analysis results for each project. Table 4.22 presents the income statement for each customer using a traditional volume-based allocation while Table 4.23 shows the income statement using ABC.

Table 4.24 shows how much the net profit percentage for each customer changes when the overhead allocation method changes to ABC. As shown in Table 4.24, net profit percentage on City of xx changed increased by 0.91%. As in the case of profitability analysis of market sectors, you may want to see profitability analysis over a longer period for strategic planning purposes.

4.6   Chapter summary

General overhead costs are usually either project management-supporting or organization-sustaining. This chapter discussed a case where ABC was applied to managing general overhead costs. The typical types of cost objects are projects, market sectors, and customers, as seen in the case study. Accurate allocation of general overhead costs to projects enables you to achieve accurate profit analysis for each project; accurate allocation to market sectors and customers allows you to focus marketing efforts on profitable market sectors and customers.

In the next chapter, we will discuss how ABC can be applied to managing overhead costs at a fabrication shop.

References

Grieco, P. and Pilachiwski, M. (1995). *Activity-Based Costing: The Key to World Class Performance*, PT Publication Inc., 4360 North Lake Blvd, FL.

Horngren, C., Foster, G., and Datar, S. (1999). *Cost Accounting*, 10th edn, Prentice Hall, Upper Saddle River, NJ.

Jackson, B. (2011). *Design-Build Essentials*, Delmar Cengage Learning, Clifton Park, NY. 411 pp.

Kim, Y. (2002). *The Implications of a New Production System for Project Cost Control*, PhD thesis, Dept. of Civil Engineering, University of California at Berkeley, Berkeley, CA, December 2002.

U.S. Green Building Council (2013). Reference Guide for Building Design and Construction, U.S. Green Building Council, Washington, DC, 806 pp.

5 Managing Overhead Costs in a Fabrication Shop

Activity Based Costing for Construction Companies, First Edition. Yong-Woo Kim.
© 2017 John Wiley & Sons Ltd. Published 2017 by John Wiley & Sons Ltd.

Suppliers provide all the materials and services needed for building projects. The term 'supplier' encompasses a very diverse range of roles and responsibilities. The definition of a supplier can vary around materials or services. Most suppliers are manufacturing fabricators in charge of fabrication and delivery to construction job sites.

Sometimes contractors extend their role to include manufacturing and delivery in order to maintain a reliable supply of important materials. For example, some contractors have a rebar fabrication shop in-house to reliably supply fabricated rebar to their construction projects. In that case, contractors need to include all material and fabrication costs in their project costs by allocating them to projects. As a result, construction costs and profits on projects may change depending on how these costs are allocated to projects.

In this chapter, we use a case study of reinforced bar fabrication costs allocated to different projects. We will discuss how reinforced bar (rebar) fabrication costs are to be allocated to projects using activity-based costing.

5.1 Rebar supply system

Common choices for building construction projects are reinforced concrete structures and steel-frame structures. Reinforced concrete structures consist of complex components, while the structural framework of the reinforced concrete structure is made up of three major processes: formwork, rebar installation, and concrete pouring. Management of the supply chains associated with these three major processes is key to successful project delivery in construction projects (Polat and Ballard, 2005). In relation to the components, the supply system for reinforced steel bar (so-called rebar) is critical for construction in order to avoid project delays (Polat and Ballard, 2005). Schedule delays in the rebar supply chain usually impact the project's performance negatively.

Traditionally, the construction industry has preferred to deliver rebar to the construction site in large batches and fabricate (i.e., cut and bend) rebar on site before installing the fabricated and assembled rebar. The traditional rebar supply system takes up a large amount of space on site (i.e., on-site yard) leading to expensive holding costs. Due to these high holding costs and space requirements, the industry has been interested in finding new ways to fabricate rebar off-site and deliver it in small batches.

A new rebar supply system would involve an off-site prefabrication shop where the rebar is fabricated. By nature, prefabrication involves a need for close collaboration between contractors (users) and suppliers. In other words, the new method will require reliable planning and predictable workflow. This new strategy involving off-site prefabrication is sometimes called a Lean supply system, as a frequent delivery of small batches needs to be pulled from a user (i.e., contractors on site) (Figure 5.1).

Although this new supply system requires more frequent deliveries, it removes yard-space requirements and improves on-site productivity. Previous research results have shown that a rebar supply system using prefabrication reduces the need for inventory space on site and improves quality and productivity due to the off-site prefabrication (Arbulu and Ballard, 2004; Naim, *et al.*, 1999).

(a)

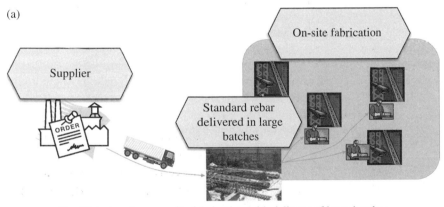

Traditional system: on-site fabrication with delivery of large batches

(b)

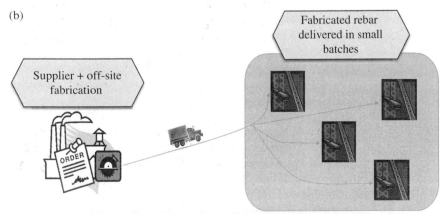

Lean supply system: prefabrication with frequent delivery of small batches

Figure 5.1   Two prevailing material supply systems of steel reinforced bar.

If the supplier is an external stakeholder, the contracted price (i.e., outsourcing cost) is transferred to a project; therefore, the outsourcing cost is included in the project's costs. If a contractor has an in-house manufacturing facility, on the other hand, those fabrication costs would be allocated to each project appropriately.

The allocation of supply costs along the supply chain has become an important question (Kong *et al.*, 2008). In the case of an in-house fabrication shop, its supply costs should be allocated to projects so that those supply costs are to be included in project costs. Depending on the way fabrication costs are allocated to projects, a project can change from being unprofitable to profitable.

Rebar fabrication costs include not only direct costs (i.e., rebar material costs and direct labor costs) but also manufacturing overhead costs. The costs of the rebar fabrication shop usually include the following cost components:

- direct material costs
- direct labor costs
- manufacturing overhead costs.

Direct material costs and labor costs are easy to track by their usage or consumption in each project. On the other hand, manufacturing overhead costs such as engineering costs are hard to track since the usage and consumption of such factors are not as easily known. Hence, we need an appropriate allocation based on allocating manufacturing overhead costs to multiple projects. In many cases, the overhead costs are allocated in proportion to the weight of the rebar (Kim *et al.*, 2011). This practice of cost allocation using the weight of rebar as an allocation base also leads to pricing practice in which rebar pricing including fabrication is based on the weight of reinforced steel bar (rebar).

This chapter examines the **cost** distortion in allocating rebar supply costs to construction **projects**. The chapter will give you two different methods: a traditional allocation method (i.e., weight-based allocation) and an activity-based costing method. We will use a case study to show how the conventional method leads to cost distortion and how ABC can contribute to process improvements.

5.2   Case study: PQR Construction Inc.[1]

This case study deals with the costs of rebar shop PQR, which supplies fabricated rebar to various construction projects. The general contractor (PQR) has practiced just-in-time (JIT) delivery and prefabrication for its rebar supply on its urban building projects, which could not accommodate the delivery of materials in large batches. In the past, the company found that the on-time delivery rate of pre-fabricated rebar was not satisfactory. The contractor's experience with pilot projects using prefabrication and JIT delivery reinforced the importance of timely delivery of rebar to avoid waiting time on site. You can imagine a situation where a contractor is waiting for rebar to be delivered with minimum rebar inventory on site.

Responding to the challenge of unreliable rebar delivery, the company acquired a rebar fabrication shop that could supply fabricated rebar to its projects. The rebar fabrication shop was a manufacturing company although it became a subsidiary of PQR construction Inc.

5.2.1   The rebar fabrication shop's cost structure

The rebar fabrication shop's cost structure is made up of direct labor costs, direct material costs, and manufacturing overhead costs. All functions of general overhead costs such as general management costs have been absorbed by its parent company, ABC Construction Inc. If the shop is an independent company, instead of a subsidiary of the construction contractor, the home office overhead costs of the rebar fabrication shop should be added to the current cost structure.

As you might expect, the company allocated all rebar fabrication costs including manufacturing overhead costs to projects so that those manufacturing costs were included in its projects' costs. Direct costs are easy to trace to each project, while manufacturing overhead costs are hard to trace or allocate. Table 5.1 shows the fabrication shop's revenue and direct costs, and the rebar usage by the projects to which the shop provided fabricated rebar.

---

[1] The chapter uses a modified case study on rebar fabrication costs (Kim *et al.*, 2011). The case study involves a simplified real project where the author worked as a consultant in the past.

Table 5.1 Cost breakdown of rebar fabrication costs.

	Project X (Commercial)	Project Y (High-Rise Condo)	Project Z (Heavy Civil)	Total
Sales revenue	$248,400.00	$210,600.00	$345,600.00	$804,600.00
Direct costs	$207,460.00	$176,670.00	$287,360.00	$671,490.00
Direct material	$193,200.00	$163,800.00	$268,800.00	$625,800.00
Direct labor	$14,260.00	$12,870.00	$18,560.00	$45,690.00
Rebar usage (tons)	460	390	640	1,490

5.2.2   *Allocation of rebar fabrication shop's costs to projects*

The company allocated its rebar fabrication shop costs to their projects using a traditional volume-based cost allocation method. A new operations manager in charge of heavy civil construction projects claimed that heavy civil projects were overcharged when rebar fabrication costs were allocated to projects. The operations manager asserted that, as compared to building projects, heavy civil projects used more standardized rebar and required less coordination efforts with design and engineering staff of the fabrication shop. Under these circumstances, heavy civil projects tended to be overcharged using volume-based allocation (i.e., rebar tons).

In responding to the operation manager's concerns about cost allocation, the company decided to apply ABC in managing their manufacturing overhead costs in order to fairly allocate manufacturing overhead costs to projects. The company formed a task force to implement ABC on their rebar fabrication shop. The task force was to explore the effects of ABC implementation on their cost allocation. The team was also asked to investigate whether, and if so to what extent, the traditional volume-based allocation method generated misleading cost information. Note that ABC implementation was initiated by the parent organization (i.e., a construction contractor), which was a major beneficiary of the fabrication shop's ABC system. Eventually a new costing system would enable the contractor to better measure profits through accurate allocation of rebar fabrication costs. The next sections present the two different allocation methods in greater detail.

5.3   Analysis using traditional rebar costs allocation

It is straightforward for the rebar fabrication shop to track its rebar costs (material costs) and labor costs to projects to which rebar is delivered. However, it is challenging to track its manufacturing overhead costs to projects. Instead of tracking overhead costs to projects, the shop should allocate manufacturing overhead costs to projects using an appropriate allocation base. Traditionally the fabrication shop allocated its manufacturing

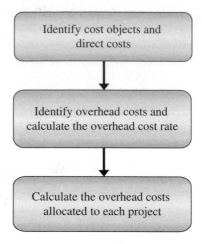

Figure 5.2   Traditional overhead allocation method procedures.

overhead costs to projects based on the usage of rebar or the number of tons delivered to each project. This section describes the procedures of overhead cost allocation using volume-based allocation, as shown in Figure 5.2.

5.3.1 Identify cost objects and direct costs

The first task is to identify cost objects and compile your direct costs. The rebar fabricator provided its fabricated rebar to three projects (X, Y, and Z). Projects X and Y were building construction projects: Project X was a commercial construction project and Project Y was a high-rise condominium project. Project Z was a heavy civil project. The case study used three cost objects, which were three projects (X, Y, and Z) to which the rebar fabrication shop supplied its fabricated rebar. Total revenue and total direct costs for one month were $804,600 and $759,790, respectively (Table 5.2).

5.3.2 Identify the overhead costs to be allocated and calculate the allocation base

The manufacturing overhead costs were $88,300, all of which was to be allocated to three projects. As mentioned, the rebar fabrication shop employed rebar usage (i.e., rebar tons fabricated and delivered) as the only allocation base to link all manufacturing

Table 5.2 Schedule of profitability using traditional costing method.

	Project X	Project Y	Project Z	Total
	(Commercial)	(High-Rise Condo)	(Heavy Civil)	
A) Sales revenue	$248,400.00	$210,600.00	$345,600.00	$804,600.00
B) Costs of goods sold	$234,720.40	$199,782.08	$325,287.52	$759,790.00
Direct material	$193,200.00	$163,800.00	$268,800.00	$625,800.00
Direct labor	$14,260.00	$12,870.00	$18,560.00	$45,690.00
Manufacturing overhead	$27,260.40	$23,112.08	$37,927.52	$88,300.00
C) Gross margin	$13,679.60	$10,817.92	$20,312.48	$44,810.00
D) Gross margin rate (C/A)	5.51%	5.14%	5.88%	5.57%

overhead costs to projects. The assumption of overhead allocation using rebar usage was that the shop's manufacturing overhead
costs (e.g., salaries of production engineers) are in proportion to
the rebar tons fabricated. For one month, as shown in Table 5.1,
the delivery record showed that total rebar tons was 460 tons in
Project X, 390 tons in Project Y, and 640 tons in Project Z.

The overhead cost rate can be determined by dividing total
manufacturing overhead costs by total quantity of the fabricated
rebar. The unit overhead cost (i.e., overhead cost per ton) can be
calculated as follows:

$$\text{Overhead cost rate} = \frac{Total\ Manufacturing\ Overhead\ Costs}{Total\ Rebar\ Tonnage}$$

$$= \frac{\$88,300}{1490\ ton}$$

5.3.3   Calculate the overhead costs allocated to each project

You need to obtain total manufacturing overhead costs from the
accounting department. Then, you can allocate total manufacturing
overhead costs in proportion to the rebar usage of each project. Note
that rebar usage (# of tons) is the allocation base in the traditional
costing. In this case, total manufacturing overhead costs ($88,300)
would be allocated to three projects (X, Y, and Z) based on the rebar
usage of each project (Table 5.1). The manufacturing overhead costs
to be allocated to each project are calculated as follows:

$$\text{Overhead costs allocated to Project X} = \frac{\$88,300}{1490\ ton} \cdot \times 460\ \text{tons} = \$27,260.40$$

$$\text{Overhead costs allocated to Project Y} = \frac{\$88,300}{1490\ ton} \times 390\ \text{tons} = \$23,112.08$$

$$\text{Overhead costs allocated to Project Z} = \frac{\$88,300}{1490\ ton} \times 640\ \text{tons} = \$37,927.52$$

Table 5.2 shows a schedule of profitability for each project
using this traditional allocation method. It is noted that total

manufacturing overhead costs ($88,300) was allocated to three projects in proportion to rebar usage of each project. The profitability ratios on those three projects were 5.51%, 5.14% and 5.88%, respectively. However, these values might be distorted as overhead costs do not vary in proportion to the rebar usage.

5.4 Analysis using activity-based costing

The task force was made up of four individuals, including one external facilitator. The team developed ABC procedures for managing the overhead costs of the rebar fabrication shop. The implementation procedures are summarized in Figure 5.3.

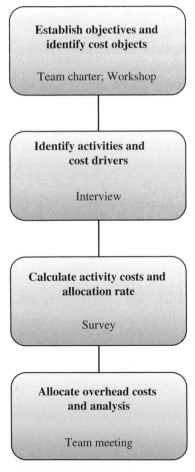

Figure 5.3   ABC overhead allocation method procedures.

5.4.1   Determining system objectives and defining cost objects

In developing an ABC system, you should first define the specific objectives of the costing system. For example, the common objectives of an ABC system are to identify major processes and determine activity costs, to measure the profitability of each cost object, and to reduce overhead costs. Well-defined system objectives enable you to set the level of activity detail, define an appropriate scope of the costing system, and select relevant cost objects. On the other hand, a poorly defined system inhibits the team from setting the appropriate level of activity detail and scope for the system and defining relevant cost objects. The more time and effort that you put into defining the system, the less time you will take in developing and executing the ABC system.

The task force developed three concrete objectives of their ABC system. The first was to investigate whether and if so how the current conventional volume-based allocation method distorts the costs and the profitability of each cost object. To this end, a comparison analysis was made after the results of the ABC system became available. The second objective was to identify which projects are profitable and which are not. The information on profitability was important because the rebar fabrication shop planned to extend its business in the near future. Lastly, accurate cost information on the projects was also needed to adjust rebar fabrication costs transferred to projects, so that projects would have more accurate cost information.

The objectives of the rebar shop's ABC system:

1) To investigate whether the current traditional volume-based allocation system distorts costs and profitability analysis.
2) To determine which projects are profitable.
3) To adjust the contract price between the rebar shop and each project so as to make the project costs more accurate.

The team also decided to exclude overhead costs related to facilities and utilities. Office rent and various utility costs were examples of overhead cost items not included in the ABC system. Instead, those overhead cost items could be allocated to projects in proportion to the rebar usage, just as a conventional volume-based allocation method would require. In summary, the ABC system in this case study included all manufacturing overhead costs excluding facility and utility costs.

Defining the objectives of the ABC system involves defining cost objects as well. While previous case studies had multiple types of cost objects, the rebar shop had a single type of cost object: the projects X, Y, and Z. Taking into account the objectives and the system scope, the task force decided to simplify activities. As in the case studies detailed in Chapters 3 and 4, developing a team charter specifying objectives is highly recommended, although it is not shown in this chapter.

5.4.2   *Identifying resources and activities*

The next step is to identify resources and activities. Activities cause resources to be consumed. In other words, activities can also be defined as business processes that incur overhead expenses. The team had to determine the level of activity detail. The expected level of detail needs to be clearly defined before data collection begins. If possible, it can be defined where a team charter is developed.

In this case study, the ABC system included only resources that incurred overhead expenses. Direct labor and materials were not included in the scope of the ABC system because they were directly tracked to projects. A list of employees in the ABC system, with their monthly wages, is to be found in Table 5.3.

The task force conducted interviews with key employees in each department and distributed surveys to every employee to identify activities to be included in the ABC system. Each department

Table 5.3   **List of human resources in ABC system.**

Resource Number	Description	Monthly rate
1	Plant manager	$11,400.00
2	Drafter 1	$7,760.00
3	Drafter 2	$7,520.00
4	Production manager	$9,800.00
5	Production engineer 1	$8,600.00
6	Production engineer 2	$7,900.00
7	QC engineer	$9,280.00
8	Secretary	$4,500.00

Table 5.4   Activity list.

Activity center	Activity number	Description
Procurement	1	Receiving orders
Engineering	2	Developing shop drawings
	3	Revising shop drawings
Inventory control	4	Receiving materials
	5	Distributing materials
Production	6	Developing production schedule
planning	7	Controlling production schedule
	8	Expediting production schedule
Quality control	9	Inspecting standard rebar
	10	Inspecting fabricated rebar

corresponded with an activity center, which had one or multiple activities as shown in Table 5.4.

5.4.3 Assigning resource costs to activities

Once activities are identified, you need to track or estimate costs for each activity. To this end, you need to implement a survey to investigate how much time each employee needs to spend on performing each activity. Generally speaking, activity costs can be estimated by multiplying the unit cost of each resource (i.e., unit wage of employees) by the volume consumed (i.e., the number of hours worked). Consider an example where three employees work in the engineering department. Each employee works 5 hrs/day on drafting shop drawings for 20 working days per month and the wage of each employee is $40/hour. Then, the monthly activity costs of drafting shop drawings would be $12,000 (5 hrs/day x 20 days x $40/hr x 3 persons). Alternatively, you can use a time–effort % method, whereby the relevant percentage of each person's time and effort is allocated to each activity.

In this project, the task force collected cost information on each department from the company's accounting division. For example, the engineering department incurred a total cost of $69,000. In order to allocate department costs to each activity, the task force used a time–effort % method for simplicity (Table 5.5). The team carried out a survey in which each employee was required to allocate the average percentage of his or her time spent on each activity in five activity centers. The result of this survey on activity costs is shown in Table 5.6.

Table 5.5 Time–effort % table.

Activity	Plant manager	Drafter 1	Drafter 2	Production manager	Production engineer 1	Production engineer 2	QC eng1	Secretary
Receiving orders	0.1				0.2			0.8
Developing shop drawings		0.6	0.7					0.1
Revising shop drawings		0.4	0.3					
Receiving materials	0.05				0.1	0.15	0.15	
Distributing materials					0.1	0.15	0.15	
Developing production schedule	0.35			0.5	0.15	0.3		
Controlling production schedule	0.3			0.3	0.25	0.25		
Expediting production schedule	0.2			0.2	0.2	0.15		
Inspecting standard rebar							0.3	0.1
Inspecting fabricated rebar							0.4	

Table 5.6   Activity costs.

Activity No.	Description	Costs
1	Receiving orders	$6,460.00
2	Developing shop drawings	$10,370.00
3	Revising shop drawings	$5,360.00
4	Receiving materials	$4,007.00
5	Distributing materials	$3,437.00
6	Developing production schedule	$12,550.00
7	Controlling production schedule	$10,485.00
8	Expediting production schedule	$7,145.00
9	Inspecting standard rebar	$3,234.00
10	Inspecting fabricated rebar	$3,712.00

5.4.4   Determining a cost driver for each activity

A cost driver is a factor that causes a change in the cost of an activity (Cokins, 1996). Identifying an appropriate cost driver for each activity is a challenge. Many practitioners advise that the best way to develop cost drivers is to actively engage the "doers" of the process (Cokins, 1996). The task force carried out interviews with key employees of the rebar fabrication shop to identify cost drivers, which are factors that are proportional to activity costs (Miller, 1992; O'Guin, 1991).

Similarly to the case studies previously outlined in the book (i.e., a project-based ABC system and a home office ABC system), the team managed to find factors meeting three criteria: (1) cause-and-effect relationship, (2) objectiveness, and (3) economic feasibility in measuring the volume of the factor.

Procurement: The only major activity in the procurement function was "receiving orders" from construction projects. Though some orders consumed more resources than others, the team decided that the factor having a positive linear correlation with total procurement costs was the number of procurement orders. The team agreed that the variance of resource consumption per order was negligible. In addition, the number of procurement orders can be tracked easily and objectively.

Engineering: Major activities in the department of engineering included developing and revising shop drawings. The team determined that appropriate allocation bases for those activities were the number of drawing sheets and the number of revised drawing sheets.

Inventory control: Major activities included receiving and distributing materials. The number of tons was used as a cost driver for the activity of "receiving materials." However, the number of distribution runs was used for the "distributing materials" activity because, in distributing materials to different production lines, a load of two tons did not take twice as many resources as a load of one ton.

Production planning: Note that the number of production runs served as the cost driver for each of the five activities: distributing production plan, developing production plan, controlling production plan, expediting production plan, and inspecting fabricated rebar. The effort required for tracking the volume of cost drivers could be reduced with a reduced number of cost drivers.

Quality control: The "inspecting standard rebar" activity used the number of tons of rebar because standard-sized rebar was delivered to the shop in bulk. On the other hand, the activity of "inspecting fabricated rebar" used the number of inspections as a cost driver.

All the cost drivers are transactional cost drivers. A list of cost drivers is to be found in Table 5.7.

Table 5.7   List of cost drivers.

Activity Number	Description	Cost driver	Type of cost driver
1	Receiving orders	Number of procurement orders	Transactional driver
2	Developing shop drawings	Number of drawing sheets	Transactional driver
3	Revising shop drawings	Number of revised drawing sheets	Transactional driver
4	Receiving materials	Number of rebar tonnage	Transactional driver
5	Distributing materials	Number of production runs	Transactional driver
6	Developing production schedule	Number of production runs	Transactional driver
7	Controlling production schedule	Number of production runs	Transactional driver
8	Expediting production schedule	Number of production runs	Transactional driver
9	Inspecting standard rebar	Number of rebar tonnage	Transactional driver
10	Inspecting fabricated rebar	Number of production runs	Transactional driver

5.4.5   *Calculating a unit rate of activity costs (cost driver rate) and allocating activity costs to cost objects*

Activity costs should be allocated to cost objects using predetermined cost drivers. The unit cost of each activity or the unit rate of activity costs can be calculated as the activity costs divided by the total volume of a cost driver. In this case study, the volume of cost drivers was acquired through document analysis since all of the cost drivers were transactional cost drivers and could be tracked only through documents. A list of cost drivers with their volumes is to be found in Table 5.8.

The unit rate of activity costs (i.e., activity costs/volume of cost driver) on activities of receiving orders and developing shop drawings, for example, would be calculated as follows:

$$\text{Allocation rate of "receiving orders"} = \frac{\$6,460}{112} = \$57.68$$

$$\text{Allocation rate of "developing shop drawings"} = \frac{\$10,370}{95} = \$109.16$$

Table 5.8   Volume of cost drivers (1 month).

Activity	Project X	Project Y	Project Z	Total volume of cost driver
Receiving orders	42	38	32	112
Developing shop drawings	44	30	21	95
Revising shop drawings	11	4	3	18
Receiving materials	460	390	640	1,490
Distributing materials	42	38	32	112
Developing production schedule	42	38	32	112
Controlling production schedule	42	38	32	112
Expediting production schedule	42	38	32	112
Inspecting standard rebar	460	390	640	1,490
Inspecting fabricated rebar	42	38	32	112

The activity costs in the engineering department allocated to each project, for example, were calculated as follows:

Shop drawing costs allocated to Project X = 44 sheets × $109.16 = $4,802.95
Shop drawing costs allocated to Project Y = 30 sheets × $109.16 = $3,274.74
Shop drawing costs allocated to Project Z = 21 sheets × $109.16 = $2,292.32
Revising drawing costs allocated to Project X = 11 sheets × $297.78 = $3,275.56
Revising drawing costs allocated to Project Y = 4 sheets × $297.78 = $1,191.11
Revising drawing costs allocated to Project Z = 3 sheets × $297.78 = $893.33

All of the activity costs allocated to each project are presented in Table 5.9.

5.5   How can ABC data be used for managerial purposes?

The three major objectives were (1) to investigate whether the current traditional volume-based allocation system distorts costs and profitability analysis, (2) to determine which projects are profitable, and (3) to adjust the rebar fabrication costs transferred to projects. To achieve these goals, the team was required to develop a schedule of profitability for each project using both the ABC system and the traditional volume-based allocation method (i.e., the number of tons of rebar), and then to compare the two. In addition to the issue of accurate allocation of manufacturing overhead costs, the team investigated how ABC data could be used for process improvements.

5.5.1   *Accurate cost information through overhead cost allocation*

The contractor was the fabrication shop's parent company. In this case study, the rebar fabrication shop did not have separate home office overhead costs. Therefore, the costs of goods sold (COGS) for the fabricated rebar consisted of direct labor costs,

Table 5.9 Overhead cost allocation using ABC.

Activity	Allocation rate	Activity cost	Project X	Project Y	Project Z
Receiving orders	$57.68	$6,460.00	$2,422.50	$2,191.79	$1,845.71
Developing shop drawings	$109.16	$10,370.00	$4,802.95	$3,274.74	$2,292.32
Revising shop drawings	$297.78	$5,360.00	$3,275.56	$1,191.11	$893.33
Receiving materials	$2.69	$4,007.00	$1,237.06	$1,048.81	$1,721.13
Distributing materials	$30.69	$3,437.00	$1,288.88	$1,166.13	$982.00
Developing production schedule	$112.05	$12,550.00	$4,706.25	$4,258.04	$3,585.71
Controlling production schedule	$93.62	$10,485.00	$3,931.88	$3,557.41	$2,995.71
Expediting production schedule	$63.79	$7,145.00	$2,679.38	$2,424.20	$2,041.43
Inspecting standard rebar	$2.17	$3,234.00	$998.42	$846.48	$1,389.10
Inspecting fabricated rebar	$33.14	$3,712.00	$1,392.00	$1,259.43	$1,060.57

direct material costs, manufacturing overhead costs, and home office overhead costs. Under the circumstances, inappropriate allocation of manufacturing overhead costs leads to distortion in costs of goods sold. Since the costs of goods sold for the fabricated rebar allocated to projects becomes part of construction costs, cost distortion in COGS leads to distortion in project costs.

The results of overhead allocation to projects using ABC provided accurate cost information leading to better understanding of the relative profitability of each project. The profitability analysis results (i.e., a schedule of income statement) are to be found in Table 5.10.

The team also developed a comparison analysis chart using Tables 5.2 and 5.10. Table 5.11 shows that Project Z (a heavy civil project) was being overcharged in the category of manufacturing overhead costs, while Projects X and Y were undercharged. Accurate manufacturing cost information can assist strategic decision-making, such as price adjustment or marketing focus. Therefore, the rebar fabrication shop's revenue or the value of individual contracts on a project could be adjusted to reflect the accurate allocation of manufacturing overhead costs. As a result, the parent company (i.e., a general contractor) and the rebar shop agreed to adjust rebar price every year based on ABC data. Accurate cost information would allow the rebar fabrication shop to focus its marketing efforts on more profitable types of projects if the rebar fabrication shop wished to consider expanding its business beyond its parent company.

5.5.2   *Cost information on processes*

A conventional allocation method uses resource-based costing or one-stage costing. In other words, cost information is provided for each resource. On the other hand, ABC using two-stage costing, which provides a process view. ABC assigns resource costs to major processes or activities before assigning costs to final cost objects. Therefore, ABC provides system users with cost information on processes (Cokins, 1996).

In this case study, the team developed a list of activities with their costs (Table 5.6). The cost information on major processes can be used in process improvement efforts when needed.

Table 5.10 Schedule of profitability using activity-based costing.

	Project X	Project Y	Project Z	Total
A) Sales revenue	$248,400.00	$210,600.00	$345,600.00	$804,600.00
B) Costs of goods sold	$240,844.79	$203,526.11	$315,419.10	$759,790.00
Direct material	$193,200.00	$163,800.00	$268,800.00	$625,800.00
Direct labor	$14,260.00	$12,870.00	$18,560.00	$45,690.00
Manufacturing overhead	$33,384.79	$26,856.11	$28,059.10	$88,300.00
Activity-based costing				
Receiving orders	$2,422.50	$2,191.79	$1,845.71	$6,460.00
Developing shop drawings	$4,802.95	$3,274.74	$2,292.32	$10,370.00
Revising shop drawings	$3,275.56	$1,191.11	$893.33	$5,360.00
Receiving materials	$1,237.06	$1,048.81	$1,721.13	$4,007.00
Distributing materials	$1,288.88	$1,166.13	$982.00	$3,437.00
Developing production schedule	$4,706.25	$4,258.04	$3,585.71	$12,550.00
Controlling production schedule	$3,931.88	$3,557.41	$2,995.71	$10,485.00
Expediting production schedule	$2,679.38	$2,424.20	$2,041.43	$7,145.00
Inspecting standard rebar	$998.42	$846.48	$1,389.10	$3,234.00
Inspecting fabricated rebar	$1,392.00	$1,259.43	$1,060.57	$3,712.00
Non-ABC				
Facility & utilities	$6,649.93	$5,637.99	$9,252.08	$21,540.00
C) Gross margin	$7,555.21	$7,073.89	$30,180.90	$44,810.00
D) Gross margin rate (C/A)	$0.03	$0.03	$0.09	$0.06

Table 5.11   Cost comparison: traditional vs. activity-based costing.

	Project X	Project Y	Project Z	Total
Traditional Costing				
Cost of goods sold	$234,720.40	$199,782.08	$325,287.52	$759,790.00
Direct material	$193,200.00	$163,800.00	$268,800.00	$625,800.00
Direct labor	$14,260.00	$12,870.00	$18,560.00	$45,690.00
Manufacturing overhead	$27,260.40	$23,112.08	$37,927.52	$88,300.00
Activity-Based Costing				
Cost of goods sold	$240,844.79	$203,526.11	$315,419.10	$759,790.00
Direct material	$193,200.00	$163,800.00	$268,800.00	$625,800.00
Direct labor	$14,260.00	$12,870.00	$18,560.00	$45,690.00
Manufacturing overhead	$33,384.79	$26,856.11	$28,059.10	$88,300.00
Difference	$6,124.39	$3,744.03	$-9,868.42	$ 0

5.5.3   *Cost driver analysis*

Your ABC system also provides information on the cost drivers of activities. The cost driver rate could be used to assess the efficiency or productivity of each process. In this regard, the cost driver information allows you to identify activities that need process improvement at the time of process reengineering by comparing the cost driver rates of activities with those of other rebar fabrication shops. The costs driver rates are presented in Table 5.12.

One construction company for which the author provided ABC consulting services used the cost driver analysis results for departmental process improvements. Each department held a monthly process improvement meeting where members discussed how they could improve the efficiency of major processes. In such meetings, the cost driver analysis results could be used for reference, assuming that the data is updated on a regular basis.

Table 5.12   Cost driver rate information.

Cost per drawing	
Developing shop drawing	$109.16
Revising shop drawing	$297.78
Cost per order	
Receiving order	$57.68
Cost per ton	
Receiving materials	$2.69
Inspecting standard rebar	$2.17
Cost per production run	
Distributing materials	$30.69
Developing production plan	$112.05
Controlling production plan	$93.62
Expediting production plan	$63.79
Inspecting fabricated rebar	$33.14

5.5.4   *Ways to reduce overhead costs*

There are two ways to reduce overhead costs in your ABC system. First, overhead costs can be reduced by each department reducing the number of cost drivers. However, reducing the number of cost drivers sometimes conflicts with operational strategy. For example, JIT-delivery strategy is not compatible with reducing the number of deliveries. Therefore, you should consider the impact of reducing the volume of a cost driver on business performance prior to taking any business improvement action.

The other way to reduce overhead costs is to reduce a cost driver rate. Reducing a cost driver rate takes process reengineering efforts, in which cost driver information can be used as a reference. Management should focus on the cost drivers which compare unfavorably to those of similar projects or business organizations.

5.6   Chapter summary

Generally speaking, manufacturing or fabrication costs such as rebar fabrication costs are transferred to the construction projects. Usually general contractors use an outsourcing strategy: they make contracts with external suppliers. Transactional costs (or contracted values) with external suppliers in charge of fabrication are transferred to construction costs. In this regard, cost distortion in fabrication has been detrimental to general contractors.

However, current management principles, such as Lean construction or supply chain management, require general contractors to build and nurture relationships with their suppliers beyond the transactional relationship. For example, reducing delivery time along supply chains involves reducing inventories on site; reduced inventory or JIT delivery requires reliable on-time delivery. One of the challenges facing general contractors is that they cannot control the production system of suppliers.

In responding to such challenges, general contractors have tried several ways to integrate suppliers' production systems into their management systems. Acquisition of external suppliers would be one method, which was the case discussed in this chapter. Although internal suppliers' costs are transferred to construction costs, a general contractor can adjust such manufacturing or fabrication costs to accurately calculate its construction costs.

This chapter looked at the example of cost distortion in the rebar fabrication shop where the heavy civil project was overcharged. The results showed that heavy civil construction projects incurred less manufacturing overhead costs per rebar ton than building construction projects because they used more standardized rebar and required less coordination efforts. The case study showed how ABC could be implemented to accurately allocate manufacturing overhead costs. Like the previous case studies, this case study also used a time–effort % method in estimating activities' costs. It also showed that the time and effort required to track the volumes of cost drivers could be reduced by using a single cost driver for multiple activities.

This case study also showed that ABC results could be used for process improvements through analysis of cost drivers. So far we have discussed various cases to which general contractors might apply ABC:

- ABC applied to managing project overhead costs (Chapter 3).
- ABC applied to managing home office overhead costs (Chapter 4).
- ABC applied to managing an internal fabrication shop's overhead costs (Chapter 5).

In the next chapter, we will discuss the implementation road-map and provide a set of guidelines so that you can begin your own ABC journey.

References

Arbulu, R. and Ballard, G. (2004). "Lean supply systems in construction," *Proceedings of the 12th Annual Conference of the International Group for Lean Construction*, 23–27 July, Oslo, Norway.

Cokins, G. (1996). *Activity-based Cost Management Making it Work: a Manager's Guide to Implementing and Sustaining an Effective ABC system*, Irwin Professional Pub., Burr Ridge, IL.

Kim, Y., Han, S., Shin, S., and Choi, K. (2011). "A case study of activity-based costing in allocating rebar fabrication costs to projects," *Construction Management and Economics*, 29 May 2011, 449–461.

Kong, W., Zhang, Q., and Song, H. (2008) "Joint cost allocation among supply chain enterprises based on cooperative game." *Proceedings of the 2008 Control and Decision Conference*, 2–4 July, China.

Miller, J. (1992). "Designing and implementing a new cost management system," *Journal of Cost Management*, Winter, 41–53.

Nairn, M., Naylor, J., and Barlow, J. (1999). "Developing lean and agile supply chains in the UK house building industry," *Proceedings of the Seventh Annual Conference, International Group for Lean Construction*, 26–28 July, Berkeley, CA.

O'Guin, M.C. (1991) *The Complete Guide to Activity-Based Costing*, Prentice Hall, New York, NY.

Polat, G. and Ballard, G. (2005) "Comparison of the economics of on-site and off-site fabrication of rebar in Turkey," *Proceedings of the 13th Annual Conference of the International Group for Lean Construction*, 19–21 July, Sydney, Australia.

6

Activity-Based Costing in Your Organization

Activity Based Costing for Construction Companies, First Edition. Yong-Woo Kim.
© 2017 John Wiley & Sons Ltd. Published 2017 by John Wiley & Sons Ltd.

In previous chapters, we have examined various cases where activity-based costing (ABC) was applied to different areas in managing overhead costs. Now, it is time to bring everything together to take on your own journey. The reason why we use the term "journey" is that activity-based costing is neither a one-time application nor a final destination. Rather than a destination, it is a vehicle for getting you up the ramp to the highway of process improvement and accurate cost allocation with refined profit information.

Your project or organization may have endless issues that are different from the cases we have covered in this book. Consequently, you cannot achieve success simply by copying the procedures described in this book. One essential piece of wisdom in relation to applying ABC to any organization is "there is no cookbook." Although there is no bible or perfect cookbook, learning common mistakes and recognizing potential challenges you may face in the journey will help you to overcome obstacles and challenges in implementing ABC in your organization or project.

This chapter will provide guidance and implementation tips for your ABC journey, set out as follows:

- The benefits of implementing ABC
- Implementation roadmap for ABC
  - Advice on major processes
  - Gate checklist
- Common mistakes in implementation

6.1   The benefits of the ABC journey

There are benefits to be obtained not only from the data analysis aspect of ABC but also from the process of implementation itself. Some general benefits of implementing ABC are:

Making people across departments cost-conscious and accountable for their processes
In general, cost accounting activities such as tracking the costs of cost objects are a segregated process, usually carried out by the accounting department. Activity-based costing (ABC), on the other hand, takes the efforts of employees across departments. The process of cost tracking or data collection requiring collective efforts involves people across departmental boundaries. In addition, each employee is aware of the cost impact of each

activity he or she performs. In this regard, the process of ABC development itself makes employees who participate in the process cost-conscious and accountable for their jobs.

Transparency of cost allocation

ABC uses a cost driver to allocate activity costs to cost objects. A cost driver is a factor that has a cause-and-effect relationship between costs and a non-financial factor. As opposed to volume-based allocation, ABC allocates activity costs to cost objects using a cost driver, thereby making the relationship between activity costs and cost objects clear and transparent. In this regard, the process of cost tracking and allocation can be said to be transparent.

Flexible costing system

Unlike the traditional costing system, the ABC system is flexible in that it may have multiple cost objects depending on the objectives of the system. You can develop various cost–benefit analyses or profitability analyses by using cost information on various cost objects.

You can use data or results from ABC as information for strategic decision-making or strategic planning. Most benefits of the ABC system as a tool for a strategic decision-making are based on the fact that ABC gives more accurate cost information on various cost objects. The following sections list examples of some of the benefits where ABC is used as a tool for strategic decision-making or planning.

Bidding with reliable cost data

With the availability of accurate information on home office overhead costs and project overhead costs, you can make your bidding numbers more competitive. In most cases, construction companies have used historical information or the predetermined percentage of direct costs based on aggregate and unreliable overhead cost information.

For example, most construction companies use the same overhead cost ratio (i.e., overhead costs/total direct costs) regardless of the type of the project they are bidding on. However, construction companies using the ABC system are likely to have more accurate overhead cost data especially with regard to different project types, such as new building construction projects vs. renovation construction projects. Accurate overhead cost information keeps construction companies more competitive in the competitive construction market.

Evaluation of subcontractors

One of benefits of a project-based ABC system is that ABC data allows you to evaluate subcontractors based on how much of your (i.e., the general contractor's) resources are consumed by each subcontractor. This claim assumes that subcontractors consuming more of a general contractor's overhead resources show better performances than do subcontractors consuming less.

With cost information on work divisions or specialty contractors (project-based ABC), you can develop a subcontractor evaluation system. If you have accumulated cost data over time, you can use that information in selecting a subcontractor for a new project. For this purpose, this book introduces a new MBR index (management burden ratio).[1] A general contractor can develop MBR for each subcontractor, which can then be used to evaluate their performance.

Note that MBR is only one of the available metrics for measuring the performance of subcontractors. There are many other important factors a general contractor can use to evaluate its subcontractors' performance. Examples include:

- On-time delivery rate (if a subcontractor supplies materials to sites)
- Percent plan completion (PPC)[2]
- Cost performance (final costs/contracted amount).[3]

Marketing strategy

These days, negotiated contracts have become popular in private markets as opposed to low-bid procurement systems where bids with the lowest price are awarded. Negotiated contracts usually require general contractors to build and nurture relationships with potential customers. As a result, general contractors' marketing costs have increased. Since general contractors want to reduce marketing costs or make marketing efforts productive, they may want to focus on profitable market sectors (or project types) and customers. If you are a general

---

[1] Refer to Chapter 3.4 and Equation 3.3

$$\left(\frac{\textit{Project Overhead Costs Allocated to a Subcontractor}}{\textit{Total Subcontracted Amount with a Subcontractor}} \right)$$

[2] PPC = the number of assignments completed on the day stated/total number of assignments made for the week (Ballard, 2000; Ballard, 1994).

[3] Cost performance of each subcontractor = (contracted amount + change order amount)/contracted amount.

contractor, how can you identify profitable market sectors and customers without accurate profit data?

ABC can help you with this type of strategic decision. If you apply ABC to your home office, you will have the types of projects and customers as your cost objects. You will also have cost information and profitability analysis data on the projects and customers. You can focus your marketing efforts on the most profitable types of projects and customers.

ABC data can also be used at an operational level. You may use cost driver information in process reengineering. The following are examples of benefits where ABC is used as a process improvement tool at operational level.

Identifying major activities
Each department needs to identify major processes if it is committed to developing an ABC system. In the process of identifying activities, team members come to better understand activities throughout their departments, e.g., what other team members do, what information other activities require, how information on other activities is processed. In many cases, employees do not understand the nature of the jobs that others do. Note that many processes are interdependent, and understanding what others do is critical in improving processes in a department. To facilitate process improvement and organizational learning, each member of the department should understand what others do.

In this regard, the process of identifying activities[4] requires each member to understand the nature of other employees' activities. I would recommend that the definition of each activity in the activity list be addressed to foster and promote organizational learning and process improvement.

Identifying the critical activities that need to be improved
The cost driver rates, or the unit rate of activity costs, can be used as a measure of process efficiency or productivity. When the rates are compared to the cost driver information for other projects or organizations, you may find activities that need to be improved.

Setting up a cost target for major processes
I observed that some organizations use the cost driver information as a reference to set up their cost target for process improvement.

---

[4] Refer to Chapter 2.4 for more information on identifying activities.

One of the construction companies that I consulted with in the past had departmental meetings on a regular basis in which team members discussed how they could improve efficiency or productivity on major processes. In those meetings, they used to set a cost target for critical processes, for example, 5% below the current cost driver rate.

6.2   Implementation roadmap for ABC

ABC is a journey, not a destination. ABC itself does not improve processes nor reduce overhead costs. Implementing ABC paves the way for process improvement and refined cost information. At this point, the book presents two implementation roadmaps. The first, more abstract, is a concept-level roadmap for an organization striving to make ABC a sustainable method with the aim of making the organization better and more competitive. The second is a more detailed implementation roadmap for planning and executing a demonstration project or a focused ABC application, one of the milestones along the on-ramp to the ABC highway.

6.2.1   Concept-level roadmap

In the first journey, you should limit the scope of the ABC system based on your urgent needs, i.e., a focused application. This focused application can be a demonstration project, the results of which may prompt you either to expand the ABC application or not to proceed further with it. In each application, an organization should go through three stages: a planning stage, an execution stage, and an internalizing stage. Although case studies in the previous chapters focus on planning and executing the ABC system, you should not neglect or ignore the stage of internalization. You need to stabilize the ABC system to prevent your organization from slipping back into the traditional costing method. Each stage will be addressed in a detailed implementation roadmap.

Once you have success with a focused application, you can extend ABC to other parts of the organization (expanded stage). For example, you can expand the ABC system to managing project overhead costs in heavy civil projects if your demonstration project was focused on managing project overhead costs in

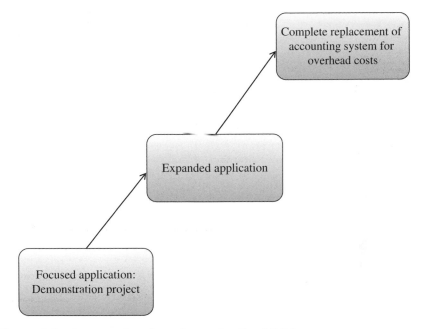

Figure 6.1   Concept-level road map for the ABC journey.

building projects. You may want to expand ABC to managing your fabrication shop if your demonstration project was focused on managing general overhead costs in your home office. As you see in the examples, you may implement multiple applications step by step in this phase.

Finally, you can replace your accounting system with ABC in managing overhead costs. Not every organization using the ABC system replaces its accounting system with an ABC system. Many organizations keep the ABC system as a complementary costing method for managerial and strategic purposes. Figure 6.1 shows a concept-level roadmap for the ABC journey.

6.2.2   Implementation roadmap for a focused application

In this book, we concentrate on focused applications. In the focused ABC application, an implementation roadmap is developed. The implementation roadmap consists of three phases (a planning stage, an execution stage, and an internalization stage) and two gates (Gate 1 and Gate 2) as shown in Figure 6.2.

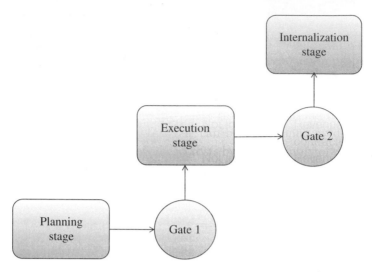

Figure 6.2   Implementation road map for a focused ABC application.

6.2.3   Phase 1. Planning stage: preparing for your ABC journey

The planning stage is the first stage in developing the ABC system. This stage helps you and your organization prepare for implementing an ABC system. The planning stage is the challenging phase, but it is easy to neglect its importance because the impact of the performance at this stage is not instantly apparent. The impact of performance at the planning stage will not be revealed until the execution stage. Please be aware that the more effort you put into this planning stage, the more time you will save in the execution and internalization stages. The planning stage involves several tasks which include the following:

- Define goals
- Build up commitment from top management
- Form a task force
- Develop a team charter in which clearly defined objectives and scope are addressed
- Educate people in preparation for ABC system development.

6.2.3.1   Prepare a clear answer to the question "Why am I planning to implement ABC?"

Before you start, you need to understand why you want to implement activity-based costing in your organization. I do not believe that you want to implement ABC simply because competitors are using it or because ABC is a popular costing method in other industries. You may have problems to solve, and you hope that ABC can help you solve them. What are those problems? Are you confident that ABC can solve the problems you have identified? If necessary, you can bring management consultants in to discuss the issues and possible outcomes.

Some of the problems or issues that ABC can help to solve or improve are:

Cost distortion, which can be an issue if the overhead costs of your organization (your home office, a project, or a manufacturing shop) are allocated to multiple cost objects;

Identification of critical activities prior to process improvement activities. You need to be aware of the costs of each activity in order to improve your processes.

In previous chapters we discussed several case studies, each of which had specific objectives for the ABC system:

- ABC in the construction project
  1) To monitor where and why project overhead costs are spent
  2) To evaluate the performance of subcontractors.
- ABC in the home office
  1) To estimate accurate profit for each project and each market sector
  2) To identify major processes in each department.
- ABC in the rebar fabrication shop
  1) To investigate whether the traditional volume-based allocation system distorts costs and profitability analysis
  2) To determine which projects are profitable
  3) To adjust the contract price between the rebar shop and each project so as to make the project cost transferred accurate.

Depending on the objectives or goals, you will have a clear idea of the scope of the system. For example, it is not a good idea to implement ABC company-wide from the beginning. In many cases, applying ABC to one or two specific department(s) is enough to achieve your goals.

6.2.3.2 Educate your people

If it is your first time implementing ABC, you may need external help. In many cases, an external consultant is brought in to lead the implementation. However, I have also observed that the skills and knowledge needed to implement ABC are not always transferred to the recipient organization. As a result, implementing ABC can be a one-time event, which may provide some improvement.

If you want to make ABC a sustainable process for improving your organization, I would advise the use of a sensei instead of a management consultant to lead the ABC task force. How is a sensei different from a management consultant? The word "sensei" is used in Japan to refer to a teacher who has mastered the subject (Liker, 2003). You need a person who can not only lead or facilitate the implementation of ABC but who can also teach the skills and knowledge to transplant ABC into the organization. You can ask a management consultant to serve as sensei before starting the process. Eventually you need to build up ABC champions in your organization who can lead and coach other people.

Leveraging champions helps convince other members of your organization to participate in ABC development and support the task force. It is my observation that people tend to resist a new management movement such as ABC when an external specialist leads it, as opposed to an internal staff member.

6.2.3.3   Form a task force and develop a team charter

You need to develop a task force to lead the implementation. The team must include members across departmental boundaries. Developing an ABC system requires a team to understand the internal processes of each department; the team must also be able to obtain resource consumption information. With team members from across departments, such information is easily accessible. It is recommended to have a team leader or external facilitator who can coordinate or lead the implementation. In many cases, a consultant or sensei takes the role of facilitator in a task force.

Once you have a task force in place, the team needs to develop a team charter (Figure 6.3), for several reasons. First, the team charter is a game plan in which the objectives (what) and methods (how) are described. Second, the process of developing the team charter helps improve the collaborative spirit required for the process of ABC development.

The team charter needs to be developed by *collaborative efforts*. It also needs to be agreed and approved by every team member. The charter without consensus may create confusion and cacophony that inhibit the team from moving forward. Although variations are allowed, a typical ABC team charter includes the following elements:

Objectives: In many cases, top management has specific goals in mind before forming a team. Although the objectives of the system are set by top management, the team members need to interpret them and re-create objectives that are achievable with the available resources (manpower, budget, and time). I have observed that the objectives set by a top manager were so varied and general that they could not be achieved with a single implementation of ABC.

Scope: Once the objectives are agreed, the team needs to define the scope of the system, depending on the objectives. It is important to avoid making the system too ambiguous or wide-open. The scope of the system should be limited so as to achieve results relevant to the objectives.

First, the team may choose which departments are to participate in developing the system. They must consider whether applying ABC to the entire organization is necessary to achieve the objectives.

Second, the team also needs to determine the types of costs to be included. The case studies in previous chapters included only human resources in their overhead costs, and excluded facility and utility costs. Depending on the needs and objectives, facility and utility costs can be included as required.

Gate 1

Gate 1 is the first gate in ABC system development and follows the planning stage. A checklist can be used for self-assessment. A checklist is a valuable tool to reduce the potential of overlooking an issue (or issues) at the planning stage. A checklist also serves as a reminder to the task force members of the required actions in the planning stage. Passing through Gate 1 allows the team into the execution stage. The checklist for Gate 1 consists of four areas, each of which has a set of information that should be checked off before moving to the next stage.

- Commitment from top management
- Task team formation
- Team charter
- Preparing the organization.

The complete checklist for Gate 1 is to be found in Figure 6.4.

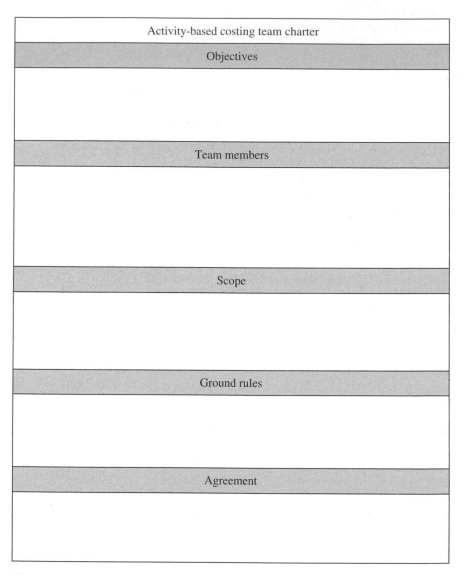

Figure 6.3   **Template form for a team charter.**

6.2.4   *Phase 2. Execution stage: developing your ABC system*

The execution stage involves implementing plans (e.g., data collection plan) created during the planning stage. The execution stage is typically the longest phase of the ABC development process. It is the stage within which the deliverables are presented to the users of the system. While plans are executed, a series of management

---

Commitment from Top Management

1. Did top management declare their commitment to the ABC system?

2. Are your employees aware of the commitment to the ABC system from top management?

Task Team Formation

1. Did you make a task force that included an external facilitator and members from

departments?

2. Did your task force have authority to access activtiy data or to collect activity

data?

Team Charter

1. Did your task force develop specific objectives for the ABC system?

2. Did your task force define the specific scope of the ABC system?

3. Did your task force determine the level of activity detail?

4. Did your task force determine the method of estimating the costs of activities?

5. Did you develop a team charter and have it endorsed by each member and a senior-level

manager in charge?

Getting your Organization Prepared

1. Did you provide training such as a workshop on ABC?

2. Are your employees aware of the obejctives of the ABC system?

3. Are your employees aware of the procedures for data collection?

Figure 6.4   A checklist for Gate 1.

processes should be undertaken to encourage employees to participate in data collection and to support ABC development.

Leadership and commitment from top management are needed to successfully implement plans in this stage. Without leadership and commitment from top management, employees tend to be reluctant to participate in ABC development processes.

In the execution stage, the task force needs to carry out various tasks, including the following:

- Define cost objects
- Develop a list of activities
- Develop a list of cost drivers
- Allocate resource costs to activities
- Calculate the unit rates of activities
- Calculate the costs of cost objects.

6.2.4.1 Reduce the number of types of cost objects

You need to define appropriate cost objects, the costs of which should provide relevant information for achieving the objectives of your ABC system. I would recommend not defining more cost objects than needed. Remember that each cost object needs a cost driver and you need to keep track of the volume of each driver on a regular basis. In the case studies discussed in previous chapters, the specific objectives of each ABC system were the following:

- **ABC in the construction project**
  Management areas (cost management, quality management, time management, etc.)
  Buildings (building #1, building #2, etc.)
  Work divisions (earthwork, structure, finishes, cladding, mechanical, electrical, etc.)
- **ABC in the home office**
  Projects (project #1, project #2, etc.)
  Market sectors (commercial, tenant improvement, heavy civil, etc.)
  Customers (customer #1, customer #2, etc.)
- **ABC in the rebar fabrication shop**
  Projects (project #1, project #2, etc.)

The task force may simulate the cost analysis with pseudo cost data prior to collecting cost information on activities. The simulation will show how each data type can be used in the analysis; the team will be able to determine if such cost objects are needed. If this is your first time implementing ABC, do not define complex cost objects. Begin with a simple structure of cost objects.

6.2.4.2 Develop a hierarchical activity list

Care should be taken to use an appropriate level of detail when the team defines activities. The process of identifying activities is similar to the process of developing a work breakdown structure (WBS) (Project Management Institute, 2013; Fleming and Koppleman, 1994). As in a WBS, the activities in the ABC system need to be arranged in a hierarchy and constructed to allow for logical groupings. As discussed in Chapter 2, there are three levels of activities

- *Daily task level* is appropriate for process reengineering. However, the ABC system can be complex if you use activities at the daily task level. Bear in mind that each activity has its cost driver. Avoid using activities at the daily task level in the ABC system because you have to update too many cost drivers on a regular basis.
- *Activity level* is appropriate for an ABC system. You should make sure that you can manage to update the costs of activities.
- *Functional level* is appropriate for use as an activity center. Usually each department or functional unit corresponds with each activity center.

I would highly recommend the separate development of an *activity list* and *activity dictionary*, so that the process of developing the costs of activities becomes efficient. You can begin to develop activity lists at the functional level. In many cases, it is straightforward to list out the activity centers first because they correspond to departments. You can then develop the list of activities for each function. When you need to improve processes for a specific activity, you can develop a list of processes associated with it. As mentioned in Chapter 4.4, you are strongly recommended to use an activity code to effectively manage activities. The first two letters to the left can be used if one function has more than one activity center. In the case of a project-based ABC system, these first two letters representing a function are seldom used. On the other hand, you might have these first two letters representing a function when you develop a home office ABC system because of its organizational complexity.

Once you have the list of activities, you need to develop an activity dictionary (Figure 6.6). In addition to a description of each activity, the activity dictionary needs to include the type of resources required

Function   Activity   Activity
(Optional)   center

to perform it. The resource information is needed when you develop a survey to estimate the costs of activities. The progression of the list of activities and activity dictionary development is as follows:

Figures 6.5 and 6.6 show samples of the list of activities and activity dictionary for your reference.

Activity center ID	Activity center	Activity ID	Activity

Figure 6.5   Template activity list.

Activity ID	Activity	Definition	Resources

Figure 6.6   Template activity dictionary.

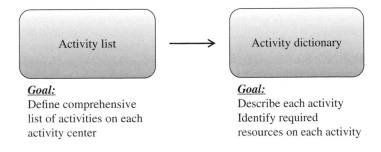

Goal:
Define comprehensive
list of activities on each
activity center

Goal:
Describe each activity
Identify required
resources on each activity

6.2.4.3   Develop a list of cost drivers meeting three criteria

After activities are identified, you need to develop a cost driver on each activity. As shown in earlier chapters, you can define a list of cost drivers after activity costs are calculated. It is acceptable to switch the sequence as long as cost drivers are defined before activity costs are assigned to cost objects.

The reason why you are recommended to define a cost driver on each activity is that the collective efforts of the task force members are required. As discussed, you need to select an appropriate cost driver on each activity from the three types of cost drivers using the following procedures:

1) The team needs to select one of the three types of cost drivers (i.e., transactional, duration, and budget) based on the attributes of each activity.
2) Once the type of cost driver is determined, a few candidate cost drivers can be developed for use as cost driver for each activity.
3) Then, the team can use the three criteria previously discussed to determine the cost driver:
    • Does it have a cause–effect relationship with a cost object?
    • Can we measure the volume of cost driver in an objective way?
    • Can we measure the volume of the cost driver in an economically feasible way?

Sometimes the team may have a hard time choosing the right cost driver from two candidates. Suppose that one gives more accurate costs while the other is simpler. One piece of practical advice from my experience as a consultant is that simplicity is paramount. This advice particularly applies if you are a first time user. Remember that you need to keep track of the volume of each cost driver; measuring it is not a one-time event.

6.2.4.4   Determine resource costs to activities

If you are a first-time implementer, I would advise you to use a time–effort % method where a percentage of each person's time is allocated to each activity. Prior to carrying out a survey to obtain this data, there are two requirements: (1) obtaining the unit costs of resources and (2) educating people participating the survey.

6.2.4.4.1 *Obtaining the unit costs of resources*

Most resources in an ABC system are human resources, so you need to obtain salary or hourly wage information on each human resource from the accounting department. The wage information usually includes burdens (taxes). When wage information outside organizational boundaries is needed, it must be obtained from a task force member. In this regard, obtaining commitment from all participating units is critical.

6.2.4.4.2 *Educating people participating in the survey*

Because people often fear job-monitoring systems, they may be uncomfortable accounting for the time spent on the different activities they perform. As a result, people put more time towards activities that look valuable and less time towards activities that look less valuable.

It is important to reassure your employees that the new system is a way to calculate the costs of each cost object not to measure the performance of each employee. For that reason, it is advisable that you hold a workshop where the purpose and major procedures of the ABC project are addressed. Alternatively, you can use an online workshop to save resources. Such educational sessions need to take place before data collection begins.

When the two prerequisites are met, you are ready to survey for time–effort % data. Each employee in the departments to be surveyed is required to allocate the average percentage of his or her time spent on each activity (Figure 6.7).

The cost of each activity can be easily calculated using the following formula, once the time–effort % form has been filled out:

$$\text{Cost of Activity i} = \sum_{J=1}^{N}(Time\%_j \times \text{Salary or Resource Cost})$$

6.2.4.5 Calculate the unit cost of activity (activity costs per cost driver)

Since the system allocates activity costs to different cost objects using a cost driver, you need to calculate the unit cost of each activity using the following simple formula:

unit cost of an activity = activity cost/total volume of cost driver

	Wage	Act #1	Act #2	Act #3	.....	Total percentage
Resc #1		%	%	%	%	100%
Resc #2		%	%	%	%	100%
Resc #3		%	%	%	%	100%
Resc #4		%	%	%	%	100%
Resc #5		%	%	%	%	100%
Resc #6		%	%	%	%	100%
Resc #7		%	%	%	%	100%
...		%	%	%	%	100%

Figure 6.7   Time–effort % survey form.

When calculating the volume of cost drivers, you need to have different strategies depending on the type of cost driver.

Budget cost driver: You may need to obtain budget-related information including

- Material budget
- Labor budget
- Total budget

- Revenue
- Spending on material.

Duration cost driver: For duration drivers, you need to measure the duration of an activity. The simple way to measure the length of the activity is to ask the person who performs the activity. It is important to use an average duration because duration can fluctuate over time.

Transactional cost driver: This is the most common type of cost driver. However, it can be cumbersome to track the volume of this type of driver. But, never fear, most transactional cost drivers can be tracked through official documentation. For example, the number of inspections can be tracked using inspection documents such as inspection reports. It is important to measure the volume of cost drivers by taking the average of at least three-months' data.

Once you obtain the volume of each cost driver, you can calculate the unit cost of each activity, the activity cost per cost driver, using the following equation:

$$UR_i = C_k / Q_i$$

Where, k = Cost object number
    i = Activity number
    C = Cost
    UR_i = Unit rate of activity i
    Q_i = Quantity (volume) of cost driver for activity i

Gate 2
Gate 2 is the second gate in ABC system development and follows the execution stage. As for Gate 1, a checklist is provided for self-assessment. Passing through Gate 2 allows the team to proceed to the internalization stage. A checklist for Gate 2 consists of two areas, each of which has a set of information that should be checked off before moving to the internalization stage.

- Developing a list of activities
- Developing a list of cost drivers and costs of activities.

The complete checklist for Gate 2 is to be found in Figure 6.8.

Developing a list of activities

1. Were the activities identified at the level of detail defined in the planning stage?

2. Did your task force develop a hierarchical activity list?

3. Did your team develop an activity dictionary?

Developing a list of cost drivers and costs of activities

1. Does each cost driver have a cause-and-effect relationship with a cost object?

2. Can team measure the volume of each cost driver in an objective way?

3. Can team measure the volume of each cost driver in an economically feasible way?

4. Do you think that the survey on time-effort % is carried out without data distortion (i.e., candid responses)?

5. Did you calculate activity costs that matched with traditional accounting data?

6. Did you calculate the unit cost of activities?

Figure 6.8   A checklist for Gate 2.

6.2.5   *Phase 3. Internalization stage: final tune-up*

The internalization stage is the final stage of ABC development. This stage involves fine-tuning the ABC system by testing the prototype system and developing a system manual that helps the organization to update and maintain the system.

6.2.5.1   Test-run the system

Having obtained the unit cost of each activity (i.e., costs per cost driver), you need to test-run the prototype system before executing it routinely. The total volume of cost drivers can be divided into cost objects. Suppose that a "change order review" activity has a cost driver of "the number of change orders" and the total volume of the cost driver is 20 (i.e., the number of change orders issued for the last three months). You should investigate how many change orders each cost object has consumed over the last three months and ensure that the total number of change orders is still 20. By doing so, you can reduce additional efforts in tracking the volume of the change order.

You need to assign the activity costs to cost objects based on the volume of cost drivers consumed by each cost object. In the process of assigning costs to cost objects, you need to make sure the prototype system meets the following condition:

$$\sum \text{Resource Cost} = \sum \text{Activity Cost} = \sum \text{Cost of Cost Object}$$

Once you are sure that these three terms produce identical results, you need to produce a variety of cost reports such as activity cost reports. The purpose is to establish whether the results generated by the prototype system can achieve the objectives set by the task force (i.e., the objectives set out in the team charter).

The primary objective of the test-run is to identify errors or areas for improvement and to refine the prototype system. At the end of the test-run, areas for improvement should be listed if any of the following conditions occur:

a) When there is any objective of the system unfulfilled by the analysis of the results of the test-run;
b) When it was difficult to track the volume of any cost driver;
c) When any transactional cost driver showed significant fluctuation over time.

In the first case, the cost objects must be redefined or new cost objects added, so that you have the information necessary to achieve the system objectives. The worst case scenario is to adjust the system objectives. In the second case, you need to consider an alternative cost driver that allows you to manage the tracking of the volume of the cost driver. In the last case, you need to consider changing to either a duration cost driver or a budget cost driver.

6.2.5.2   Develop a system manual

When you refine the prototype system based on the finds of the test-run, you should develop a system user manual. A system user manual is a written guide in hardcopy (paper report) or electronic document format that provides instructions on how to use the ABC system. Some firms also develop educational material in web-video or other mobile format so that they can minimize time spent on class-type instruction.

Developing a system manual is a critical process because the ABC system requires regular updates. The system manual is expected to include the following components:

1) *Scope and Objectives*
2) *Activity List*
   You need to include the activity hierarchy and the definition of each activity.
3) *Resource List*
   The manual needs to clearly show what specific resources are consumed by each activity.
4) *Allocation Procedure*
   You need to delineate how the volume of cost drivers is tracked and how activity costs are allocated to cost objects.
5) *Types of Reports*
   The system manual should indicate what types of ABC cost reports need to be generated.
6) *Update Instruction on Activity Costs*
   The activity costs in the prototype may change over time. Therefore, you need to address the frequency of updating activity costs and the method for updating them. The frequency of updating activity costs may differ depending on the attributes of activities.
7) *Update Instruction on the Unit Cost of Cost Drivers*
   When the activity costs change as a result of updates, the unit cost of cost drivers should change accordingly.

6.3   Common mistakes in the journey

6.3.1   *Beginning your ABC journey without strong commitment from top management*

Commitment to the ABC system is most effective when it comes directly from the top of an organization. I have observed that commitment from top management is one of the most important success factors during ABC system development. Commitment from top management makes employees and members of organizations aware of the sense of urgency and ensures that tasks associated with the ABC system are put on their priority lists. Otherwise, the tasks associated with an ABC system can be seen as burdensome.

Commitment from top management is needed even after the ABC system is developed. Without deep commitment with consistent leadership, organizations tend to slip back into traditional cost management once operating the system gets tough, perhaps in relation to the need to update the volume of cost drivers on a regular basis. This is in line with Kotter's experience (Kotter, 1995; Kotter and Cohen, 2002) that 50% of new operational management approaches failed at this step.

6.3.2 *Beginning your journey with poorly defined objectives and scope*

The more time and effort is put into defining the objectives and scope of the system, the more time will be saved in its execution, including time required for data collection and analysis. If you jump into developing the ABC system, perhaps defining activities without properly defining objectives and scope, it will not be surprising if you are forced to redefine activities or if you get results that are not relevant to your goals.

6.3.3 *Developing a task force that does not have the necessary authority*

The first time I was asked to consult on ABC implementation for a contractor, the task force was made up of new or inexperienced employees from several departments because most departments did not take the ABC initiative seriously. Instead, they thought that it was a kind of business campaign on which they were required to spend their time and resources.

I liked working with them, but there were two major problems in implementing ABC. First, the team had difficulty developing activities, since they were not fully aware of the processes of the department in which they worked. Second, the team also had difficulty in collecting activity data such as duration or cost driver volumes, since they had limited access to such data.

> **Tip!**
>
> When you form a new task force leading ABC implementation, make sure that each team member is experienced and is aware of the major processes within their departments. In addition, you need to give authority to a task force so that they have easy access to various data.

6.3.4 *Developing more cost objects than needed*

The team may attempt to have as many cost objects as possible. However, you should keep in mind that every additional cost object needs additional efforts to keep track of its cost driver. You can avoid such a mistake by having a clear set of objectives prior to developing a team charter.

> **Tip!**
>
> You need to develop a set of specific objectives to be achieved through the implementation of ABC. Then, select the minimum number of cost objects needed to achieve the objectives.

6.3.5 *Making activities ambiguous*

When you develop activities, there should be consensus on the definition of each activity. Sometimes, the activity seems too obvious to the team members to discuss its definition and scope. However, the definition or scope of an activity can change as time passes. Two years after the system is developed, for example, people may have difficulty calculating activity costs because of poorly defined activities.

> **Tip!**
>
> You need to develop a clearly defined list of activities. When you develop a list of activities, you need to use a hierarchy to differentiate activities from functions or detailed processes.

6.3.6 *The effect of distorted time–effort % assigned to activities*

Some people tend to allocate a higher percentage of their time towards activities that seem more valuable. That is because they may construe the survey as intended to evaluate performance. When people distort their time–effort % data, you may have distorted ABC results. The time–effort % survey is so important and crucial in developing the ABC system because responses to the survey lead to establishing the costs of activities.

Tip!

You need to educate people before you begin collecting data and before carrying out the time–effort % survey. A two- or three-hour workshop led by an external consultant or sensei is advisable. You may use online tools such as YouTube or PowerPoint for those who do not have time to attend such workshops.

6.3.7 *Choosing cost drivers that are hard to measure*

When you choose a cost driver, you need to consider not only the cause-and-effect relationship but also its tractability. Suppose that you have a cost driver that has a positive correlation with the cost of an activity. The cost driver is not appropriate, however, if you have difficulty tracking the volume. In a case where you need to select a cost driver from two factors, I would advise you to select the one whose volume is easy to trace.

Tip!

If you establish a particular cost driver, remember that you have to continue to track the volume of a cost driver on a regular basis. After completing a test-run, it is better to change the cost driver to a more manageable one if you find that updating the volume of the cost driver is overly cumbersome.

Index

Activity Based Costing for Construction Companies, First Edition. Yong-Woo Kim.
© 2017 John Wiley & Sons Ltd. Published 2017 by John Wiley & Sons Ltd.